T0160434

IN DEFENSE OF PLANTS

IN DEFENSE
OF PLANTS

An *Exploration*
into the *Wonder*
of *Plants*

MATT CANDEIAS, PhD

CORAL GABLES

Published by Mango Publishing Group, a division of Mango Media Inc.

Cover, Layout & Design: Morgane Leoni
Cover Illustration: © mdlne / Adobe Stock

For permission requests, please contact the publisher at:

Mango Publishing Group
2850 S Douglas Road, 2nd Floor
Coral Gables, FL 33134 USA
info@mango.bz

For special orders, quantity sales, course adoptions and corporate sales, please
email the publisher at sales@mango.bz. For trade and wholesale sales, please
contact Ingram Publisher Services at: customer.service@ingramcontent.com
or +1.800.509.4887.

In Defense of Plants: An Exploration into the Wonder of Plants

Library of Congress Cataloging-in-Publication number: pending
ISBN: (p) 978-1-64250-453-8, (e) 978-1-64250-454-5
BISAC category code SCI011000—SCIENCE / Life Sciences / Botany

Printed in the United States of America

Contents

Why in Defense of Plants?

P lants don't really have a voice in today's busy world. In fact, most people I talk with think plants are boring. It is a sad reality that when most people step outdoors, at best, they see a static green wall. If they do notice plants at all, it is likely those that either cause them issues (e.g. poison ivy in the hedge or a rebel dandelion on the lawn) or that they have some sort of use for (e.g. lavender, corn, or bananas). Whether for economic gain or some purported medicinal benefit, we only seem to care what plants can do for humans. This is a travesty because plants are incredible living organisms that conquered land long before any animal crawled out of the ocean. They are fighting for survival just like any other form of life and their sessile nature means they are doing so in remarkable ways. If there is one thing that my efforts with the *In Defense of Plants Podcast* has taught me, it's that even the slightest familiarity with plant biology, ecology, and evolution will demonstrate that plants are far more dynamic than any of our forebearers could possibly have imagined. Sadly, these stories often go untold.

Hop online or pick up a book and you will find that a vast majority of popular plant literature focuses on two major areas of interest: folklore and herbalism. Imagine wanting to look up information on animals such as black bears or pangolins, but all you can find is how to cut them up, process their organs, and make tinctures or food with them. Any rational person would be outraged by this. These animals are so much more than just what their parts can supposedly do for us. And yet, such tales are the standard for plants. It's almost as if interest in plants peaked in some bygone era before we really understood what plants are about. Over the last century, science has revealed that plants aren't static backdrops to more charismatic life forms like birds or mammals. Plants are active players in the drama that is life. Even more, they are largely responsible for life as we know it. Every terrestrial biome on planet Earth begins with plants. Aquatic biomes aren't exempt, either. Aside from deep-sea thermal vents, aquatic systems around the globe depend on photosynthesis, whether it be from algae, sea grass, countless species of phytoplankton, or plants washed into the water from the land.

At the heart of it all is photosynthesis. This wonderful biological Rube Goldberg allows plants to capture energy from our nearest star and use it to break apart water and CO_2 gas to build complex organic molecules like sugars. Without photosynthesis, we would be living on a closed, finite planet. Hell, let's be honest—without photosynthesis, we wouldn't be here at all. Our own story is tied to plants, yet we treat them like inert tools. Some of them move at alarming speeds to capture animal prey while others wage chemical warfare below ground, unbeknownst to most of us.

My goal in writing this book is that I want you to see plants, even if for a moment, how I see plants. Maybe some of you will even be bitten by the botanical bug. Our planet certainly could benefit from more plant fanatics running around. What follows are stories about how plants have changed my perspective on the world. As you read, I ask you to keep a couple things in mind. For one, at any point in this book, you might be thinking something to the effect of "you forgot to mention this or that." Did I? Or perhaps is that simply a tale for another time? In writing a book like this, one must always remember that there are endless examples and facts to draw on from nature. It is simultaneously fascinating and daunting. This book is not meant to be a complete story, nor is it entirely autobiographical. Instead, it is an ode to my journey and experiences thus far.

The second is, while I strive for scientific accuracy in my communication constantly, this is not meant to be a textbook. A wise person once told me (I wish I could remember their name) that for science communication to be successful, the stories need room to breathe. As such, I take some liberties in my choice of words. I want to make it clear that I do not think plants are conscious in any way that we can comprehend. The rise in consciousness as a topic of discussion, I think, stems from a lack of imagination. Plants do not have brains; they do not have a nervous system. As far as we can tell, there is no central processing unit in plants. Plants operate largely through diffuse chemical signaling, and to think that any human possesses the capabilities to understand how a plant interacts with and perceives the world around it is to demonstrate a hubris that can only come from our narcissistic minds. I take some liberty in anthropomorphizing certain situations purely because I think it helps the reader connect to plants a little bit easier. At no point should you interpret that in any way other than as a convenient metaphor.

Finally, evolution does not have agency. It is not a hierarchical process. Evolution via natural selection works with what it has available to it, culling things that don't work and rewarding those that do by allowing them to live long enough to reproduce. Evolution is an unthinking and unfeeling force of nature, but that doesn't mean we have to talk about it in those ways. I would much rather someone walk away from this book appreciating that evolution occurs and has shaped all life on this planet in remarkable ways than bore the reader with hyper-specific jargon. If that bothers you, there are shelves of academic books waiting for you. What follows is a celebration of plants as the incredible organisms that they are. These pages are filled with personal discovery and scientific wonder, and it is my hope that each of you comes away thinking about plants a little bit more in your daily life. I am here to defend plants.

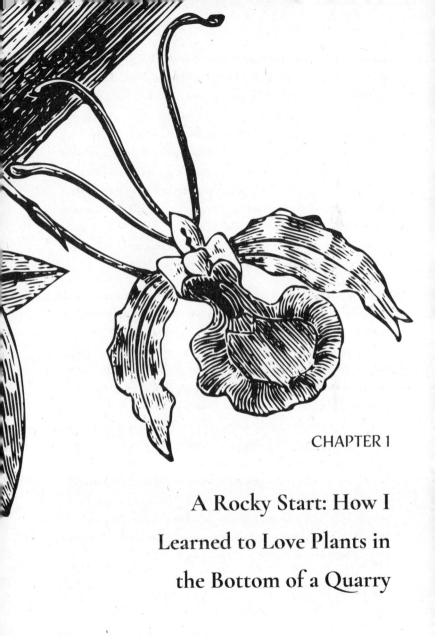

CHAPTER 1

A Rocky Start: How I
Learned to Love Plants in
the Bottom of a Quarry

I have a confession: I used to think plants were boring. I wasn't alone, either. I know for a fact that this is an opinion shared by far too many people. That's not to say I didn't love nature. Far from it. I have always been a nature nut, but my early interest in the outdoors was consumed by things that could move like insects, fish, lizards, and snakes. Fish were my first real obsession. My grandfather, who was an avid gardener, used to take me to a nearby creek that ran through town, and we would spend hours trying to catch minnows. Fish just seemed so otherworldly. They lived in a medium that I could not, and their entire anatomy was so different from what I was used to with my hands and legs. I loved the mystery that surrounded fish, and it never really left me.

By the time I got to high school, my obsession with aquarium fish was in full swing. I had amassed numerous fish tanks in my tiny bedroom, much to the chagrin of my parents and their electric bill. By this time, the hobby had taken on new meaning. I was always trying to recreate the kinds of habitats in which my fish originated. Instead of puke-colored gravel and a bubbly treasure chest, I opted for soft, sandy bottoms and lush vegetation. Growing plants in an aquarium is no joke. Aquatic plants are a unique case in the botanical world. Like whales, the ancestors of most aquatic plants started on land. As such, the adaptations they needed for survival underwater made them a lot pickier than the pothos my mom was growing on the windowsill. Growing most aquatic plants proved too difficult from my angsty teenage brain and meagre budget, so fish continued to command most of my attention. It would be a few more years before plants ever entered back into my life in any serious way.

When the time came to go off to college, I had managed to make the connection between my aquarium hobby and the natural world. I decided to major in zoology and was seriously considering a career in fisheries biology, and for a few years, at least, I really stuck to that plan. I was taking classes in subjects like invertebrate zoology, parasitology, and ichthyology and largely enjoying myself in the process. Whereas I always felt like a nerd and outcast in high school, I was finally coming into my own in college. The people around me were also nerdy, and I no longer had to stifle my urge to geek out about the natural world. For the first time in my life, I was feeling like I was moving in a direction that made sense.

Everything changed the day I took a field trip to a fishery. I never thought of myself as a squeamish person. Even the smell of roadkill was never overly offensive to me. That fishery cured me of such untested confidence. As soon as I stepped in the door, I was overwhelmed by the smell of dead fish. It was like walking into the seafood section of a supermarket after the power had been out for a week. I couldn't even finish the tour. There was no way I could work in such a place, and with that realization came the feeling that my career plans were crumpling in front of me. What was I going to do?

After lots of thinking, I decided to change gears. I transferred to a different school and changed my major from zoology to ecology. Thanks to a lengthy conversation with a friend, I realized that my love of nature stemmed more from understanding how ecosystems worked than from any individual organism. I wanted to learn how all life fits together, what drives evolution, and why we see animals in some places and not others. Ecology was the perfect subject for

A Rocky Start: How I Learned to Love Plants in the Bottom of a Quarry

me because it is the study of the interactions among organisms. The deeper I dove into the science of ecology, the more I realized I was never going to be bored again. There was so much to learn about what ecologists already knew and exponentially more to learn about aspects of the living world that were still a mystery. Plants still hadn't entered the equation yet, but ecology was broadening my horizons. I was thinking more and more about how energy moved through the environment. For instance, I learned the science behind food webs, which taught me why there are more plants than herbivores, and more herbivores than predators. Every time one organism eats another, some of the energy the prey contained is lost. As such, energy diminishes from plants to herbivore to predator. Ecology was turning on lights in my brain at the right place and at the right time in my life.

It was also around this time that I landed the job that changed my life forever. It started in a class called Restoration Ecology. The professor was a kind man with a thick Canadian accent named Dr. Chris Larson. He taught the class seminar-style, and it was largely geared toward grad students. I was one of only three undergrads in the class. Each week, we were assigned readings from a series of books. One of these was an incredible work by William K. Stevens titled *Miracle Under the Oaks: The Revival of Nature in America*. The book detailed the trials and tribulations of a grass-roots prairie restoration effort in Chicago. Now, if you are a student of any of the biological sciences, one of the overarching themes is that humans are destroying habitats. It is so pervasive that it often seems like the only truism in biology. But here was this book celebrating a group of people who had decided that habitats didn't need to disappear forever. They were

In Defense of Plants

trying to put the pieces back together. The more I learned as the semester wore on, the more I was wooed by this idea of restoration. The thought of taking damaged areas of our planet and coaxing them back into functioning ecosystems was exhilarating. I was finally learning how the theoretical principals taught in class could be applied to something real and tangible, and all of it relied on a deep understanding of the interconnectedness of nature.

As that semester was drawing to a close, I had another important realization: I needed a job. Part-time work at a pet store was not going to get me out of my parents' house. Luckily, my classmate Ashley was moving after graduation, and her bosses tasked her with finding a replacement. I inquired further and she told me that her job was based in a limestone quarry and consisted of making sure the mining company was keeping up with environmental regulations. I fancied myself something of an environmental activist at this point in my life, so the idea of working for a mining company in environmental permitting was more than a little upsetting, to say the least. However, something in my head told me I needed to know more. As we walked out of class one afternoon, she gave me a brief rundown of what she did week to week. Most of it seemed pretty standard—make sure the company isn't dumping too much sediment into local waterways, set up seismographs to make sure blasts weren't damaging neighboring residential foundations, and hand out coupons for free carwashes to neighbors who complained about all of the dust that landed on their cars. I wasn't really sold on the idea until she mentioned her side project—habitat restoration. I remember wondering why she hadn't led with that.

She told me that the company was trying to improve its image a bit by engaging in habitat restoration projects on a few of their properties. This was quite a departure from most mining operations. Usually, mining companies do one of two things with their spent quarries. They either let them fill with water and become extremely deep ponds, or, if they are shallow enough, they backfill them, plant some grass, and sell them to housing developers. This company apparently had some unique properties that provided interesting opportunities for habitat restoration. One such property was an old sand and gravel quarry located in the southern tier of western New York. The massive deposit of sand and gravel was the result of the region's glacial past. Whereas most of the surrounding soils consisted of heavy clays and rocky, glacial till, this chunk of land sat on top of a massive deposit of sand and gravel. As a result, it provided an interesting challenge for habitat restoration. Luckily, the mining company had gone outside its walls to ask for help from professional biologists, and a unique restoration plan was put into place.

The project at the sand and gravel quarry involved restoring the habitat for a tiny butterfly called the Karner blue (*Lycaeides melissa samuelis*). What this butterfly lacks in size, it makes up for in being extraordinarily beautiful. Its wings are covered in powdery blue scale that are rimmed in jet black. The bottom of each wing also sports a series of small orange crescents. You have to work to see one of these tiny butterflies in detail, but it makes the experience all the more exhilarating. Sadly, this beautiful little insect hasn't fared so well in our industrialized society. Populations of the Karner blue once ranged from coastal regions of New York and New Jersey all the way into parts of Minnesota. Today, it can only be found in a fraction of that range and is largely reduced to isolated populations. Like so many other species on our planet, its decline is largely due to habitat destruction.

Habitat destruction comes in many forms, especially for species with complex ecological needs like the Karner blue butterfly. For starters, logging, farming, and housing development fracture the landscape, creating smaller and smaller patches of suitable habitat. Those small patches of habitat become increasingly susceptible to further degradation from the encroachment of invasive species which crowd out native species. The habitats in which the Karner blue lives are also prone to fires. Although the massive wildfires occurring in places like California and Australia are horrifically destructive to people and nature alike, many ecosystems on our planet require fire to persist. The sad part is humans generally look at fire as a negative force on the landscape that must be stopped. When Europeans arrived at this continent, they set to work making sure that fires stopped happening.

Historically, fires would have burned through Karner blue habitats every few years. In doing so, they cleared the ground of a lot of woody debris and leaves. This meant that no single fire would ever have enough fuel to get out of control. Fires also kill off vegetation like shrubs and small trees that aren't adapted to cope with heat. This opens up the habitat, keeping trees widely spaced and allowing more light to reach the soil, which in turn allows plants that can handle fire to multiply in abundance. All these processes once provided ample habitat for species like the Karner blue butterfly. Today, thanks to a lack of fire, much of the Karner blue's remaining habitats have become choked with invasive plants and shrubs. Encroachment of invasive plants has changed the entire dynamic of the ecosystem and, worst of all, has pushed out a species that is particularly important for the Karner blue butterfly, perennial blue lupine (*Lupinus perennis*).

Blue lupine is a spectacular plant. At maturity, it can stand upwards of two feet (0.6m) tall and produces lovely palmately compound leaves that look like hairy green fans. When lupine goes into reproductive mode, a spike emerges from the center of the plant. This spike is covered from top to bottom with violet-blue pea-like flowers that attract insects from far and wide. The whole spectacle is truly a sight to behold.

Most insects on our planet are specialists. They have evolved to utilize only a handful or even a single species for feeding and breeding. The Karner blue butterfly is one such species. Its caterpillars can eat only the leaves of blue lupine. No other plants will do. Losing blue lupine on the landscape means losing Karner blue butterflies. This concept was at the core of the restoration efforts in the sand and gravel pit. Ashley didn't have to twist my arm to convince me to apply for the job. I applied and was hired a few weeks later.

Blue lupine in all its glory. Bees relish its beautiful flowers.

I was excited to finally get my hands dirty. The permitting aspects of the job were pretty straightforward and not all that exciting, so I poured most of my energy into learning about Karner blue butterfly habitat restoration. I learned that for the project to be successful, we needed to ensure that the property could support Karner blue caterpillars. The caterpillars themselves are kind of cute when you see them up close. They are small, chubby, and emerald green. They don't possess much in the way of defenses, either. They don't grow stinging hairs, they don't sequester poisons in their skin, nor do they secrete foul-smelling odors. If you were a hungry bird, they would make a nice snack. However, they aren't completely defenseless. Far from it, in fact. One of the coolest aspects of their ecology is that Karner blue caterpillars have teamed up with a few species of ant.

Anyone who has ever disturbed an ant hill quickly learns how vicious ants can be. They aggressively defend their homes as well as reliable sources of food. Through evolutionary time, the Karner blue butterfly (and many of its close relatives) have tapped into the ants' need to protect reliable food sources, striking up a symbiotic relationship. If you look near the rear end of a Karner blue caterpillar, you will find a tiny gland. That gland secretes a special liquid that is chock full of sugars, amino acids, and water. Some ants have figured out that, with a little prodding, they can coax the caterpillar to secrete that fluid. Ants learn to associate each caterpillar with a reliable source of food and teams of ants begin tending caterpillars as they feed. Anything that may threaten the caterpillar, be it a predatory insect, spider, or bird, will have to contend with a small army of ants. The ants are effective, too. Caterpillars tended by ants are sixty-seven percent more likely to survive than caterpillars that aren't. This relationship may even go beyond defense. Ants live in high densities and are very susceptible to attacks from fungi and other microbes. To protect themselves, ants, too, have a tiny gland on their body called metapleural gland, which secretes antimicrobial fluids that the ants scrub all over their bodies. It turns out that some of this fluid is also transferred to the caterpillars as the ants dote on them. As such, ants may also help Karner blue caterpillars fight disease.

It is funny to me how most nature programs focus on animals from places far away. They act as if the only interesting things in nature happen deep in the jungles or out on the African savanna. What I was learning was that so many incredible ecological interactions were happening right in my own backyard. I found that my environmental interests were no longer drifting away from home but instead were being consumed by things going on in my own neck of the woods. Learning about these fascinating ecological relationships made me fall head over heels for this project. I wanted to do everything in my power to make sure that the restoration efforts would succeed. However, it wasn't long before an even weirder thought started creeping into my mind. I was beginning to understand that none of these amazing relationships would be possible if it were not for blue lupine. These plants were the hub for all these remarkable interactions. If I wanted to help this project to succeed, I had to keep that in mind.

When it finally came time to visit the site, I was blown away by the progress that had already been made. I rolled into the parking lot one early spring morning in 2008 and was greeted by a tall bearded man. His name was Mike Meyers, but he bore absolutely no resemblance to the murderous villain of the *Halloween* franchise. Instead, he was soft spoken and gentle, with an inclination toward native pollinators. We made our introductions, and he instantly launched into the background of the site. It turns out the Karner blue butterfly project was largely his brainchild. Whereas the quarry supervisors were happy to provide the site and the financial backing for the restoration efforts, Mike was the person who realized the true potential of this now-defunct sand and gravel pit.

As he talked, I could tell he was growing more and more excited about the project, so we moved on from the parking lot and began surveying the restoration site. We hiked up and over one of the larger berms and were greeted with an amazing vista. Down below was a small pond that was a leftover from the active quarry days. Much of it was lined in swaying cattails (genus *Typha*) and I could hear the redwing blackbirds calling for mates. Down at my feet, wild strawberries (*Fragaria virginiana*) carpeted the hillside. Mike bent down and picked a few, handed them to me, smiled, and pointed across the pond. On the berm opposite us was a wall of color. Two years earlier, that berm was seeded with a native wildflower mix full of species that didn't mind a bit of drought. Besides lacking nutrients, sandy soils don't retain water for very long. Without spongey organic material like decaying leaves and wood, water drains away nearly as fast as it falls. Anything growing in sand must be okay with its roots drying out. Mike proceeded to tell me that not only had the berm proven that seed sowing could work, it also had won them a pollinator habitat award and they were very excited to expand the application. As he waxed poetic about the process, I began noticing signs of life all around me.

Bees and butterflies were flitting around, visiting what flowers were in bloom. On the bare ground around us, I could make out the tracks of either a fox or a coyote that obviously enjoyed the vantage point the elevated terrain provided. We walked down to the edge of the pond and it was wriggling with toad tadpoles. The idealistic environmentalist in me was dumbfounded. Open pit mining is one of the most destructive practices that has ever been invented. To gain access to whatever geological layer they are interested in, mining companies first strip away all the topsoil covering it. With the topsoil goes all the life that it once sustained: plants, microbes, insects, everything. Whatever life once called this place home had been completely removed. Yet, here we were, staring at a preponderance of life that rivaled most of the surrounding landscape. Don't get me wrong, I am by no means excusing the destruction of wild places. I firmly believe that, if our species has any hope, we need to protect wild spaces at all costs. However, this quarry was firsthand proof that humans could also put at least some of the pieces of our damaged planet back together.

The dominant vegetation growing around us largely consisted of strange-looking grasses. These were growing in dense clumps with open spaces in between. Those open spaces boasted beautiful wildflowers whose identities were also a complete mystery. Being the inquisitive observer, I asked Mike why this place looked so strange. He proceeded to tell me that, thanks to the drought-prone nature of the sandy soils, they couldn't settle with using more familiar, cool season grasses. Instead, they opted to establish a meadow comprised of warm season grasses, the likes of which one doesn't frequently encounter in cooler, wetter regions like western New York. One was a tall, lanky species with blue-green leaves and golden-brown seed heads called Indian grass (*Sorghastrum nutans*), and the other was a small, scrappy looking plant with hints of blue and red throughout called little bluestem (*Schizachyrium scoparium*). Both species were completely new to me, and I was feeling very stupid at that point. Here I was, freshly hired to undertake a massive part of this restoration project, and I had never come across the terms "cool season" or "warm season" grasses before. Mike laughed and kindly provided me a quick lesson on grass biology.

Plants are constantly having to balance the exchange of CO_2 and oxygen with the loss of water. They do so using tiny pores on their leaves and stems called stomata. When temperatures rise, causing water to evaporate from the stomata at a higher rate, plants can close those pores to avoid dehydrating. However, in doing so, they also cut off the supply of CO_2 coming into the leaf while at the same time allowing oxygen concentrations to build. The buildup of oxygen becomes a serious problem for plants, thanks to the activities of a vitally important enzyme called RuBisCO. RuBisCO's job is to take carbon molecules from CO_2 and help build sugars with them. Not only is RuBisCO the most abundant enzyme on our planet, its role in photosynthesis also makes it one of the most important. However, RuBisCO can be surprisingly bad at its job.

Little blue stem demonstrating just how beautiful grasses can be.

Nearly half of the time, RuBisCO grabs oxygen instead of CO_2. This is bad news for plants because it leads to the formation of toxins, and plants must spend a lot of energy purging them from their cells. Many billions of years ago, when photosynthesis first evolved, Earth's atmosphere contained far more CO_2 than it does today, which meant RuBisCO always had plenty of CO_2 available. Today, thanks entirely to the success of photosynthetic organisms, our atmosphere now contains way more oxygen than it used to. Unfortunately, by the time rising oxygen levels became an issue, RuBisCO was already *the* enzyme used by photosynthetic organisms.

So, how does all this relate to grasses? Well, because photosynthesis produces oxygen as a byproduct, there is always some of it floating around inside the leaves where RuBisCO can grab hold. Warm season grasses have largely gotten around this issue thanks to the evolution of some specialized photosynthetic anatomy. Their photosynthetic machinery is concentrated into dense rings of cells called bundle-sheaths that surround the leaf veins. Also, instead of using CO_2 gas directly, warm season grasses turn the gas into a couple different kinds of organic acids. These acids are then shuttled into the bundle-sheath cells where photosynthesis takes place. You can think of the process as a botanical version of how a turbo charger forces oxygen into the combustion chamber of an engine to give it more power. By obtaining carbon this way, warm season grasses can close their stomata during the heat of the day and still have plenty of carbon available for photosynthesis, which means they can remain active for longer as summer warms up and not have to deal with a rapid buildup of oxygen. In fact, this is such a successful way to deal with dry conditions that it is estimated that this type of photosynthetic anatomy has independently evolved in nineteen different plant families around the world.

Cool season grasses, on the other hand, do not possess such specialized photosynthetic anatomy. Their photosynthetic machinery is spread throughout the entire leaf with no cellular turbo chargers to concentrate carbon for RuBisCO to use. If cool season grasses want to make food, they must keep their stomata open to keep CO_2 concentrations high. Keeping those pores open also means losing valuable water, which only gets worse as temperatures increase. As the cool season grasses begin to dehydrate, they close their stomata, cutting off the supply of CO_2. Unfortunately, because they don't stop photosynthesizing, the ratio of oxygen to CO_2 within the leaf quickly increases, leading to a buildup of those toxic compounds I mentioned earlier.

Such drastic differences in the photosynthetic anatomy of warm and cool season grasses means that cool season grasses are easily outcompeted in warmer, drier climates. And, because of their ability to withstand drought, warm season grasses were therefore the perfect candidates to jump start the restoration process in this hot, dry pit of sand and gravel. Looking around, I could see the process had worked amazingly well.

As we walked, I was growing a bit impatient. I had heard so much about the Karner blue's relationship with lupine that I needed to see the plant for myself. I asked Mike to show me one, so we walked over to a corner of the property and growing amidst a sea of grasses

A Rocky Start: How I Learned to Love Plants in the Bottom of a Quarry

was a single lupine. We were too late to see any flowers but in the middle of the plant stood a rigid stem covered in hairy pea pods. Some of the pods looked as if they had exploded as each side of the pod was curled back in a violent manner. Mike told me this was how the plant dispersed its seeds. As the seeds within mature, the walls of the pod gradually dry out, creating lots of tension. With an audible pop, the pods eventually explode, catapulting the seeds out into the environment. This ballistic form of seed dispersal ensures that most of the seeds are flung away from the parent plant.

I started searching all around this massive individual to see if I could find any baby lupine. After a few minutes, I gave up. It didn't appear to be reproducing in this spot. Mike told me that this is where I would come in. He let me in on the history of the lupine restoration process at the site and, to be honest, it was a bit of a bummer. To date, they had not had much luck. For starters, the large plant we were observing was one of the first and only plants to establish on-site. It had been hand planted in this spot a few years earlier as a test to see if lupine could even grow here. Grow it did, but without more lupine, successful pollination was low.

Early seeding efforts at the quarry weren't very systematic, either. Mike and others had made a few attempts to scatter seeds on bare ground to no avail. Most seeds did not germinate, and those that did weren't living very long. They decided to switch gears to what they thought was a gentler process. Seeds were planted in seed trays and allowed to germinate and grow for a few weeks before being transplanted on site. Growing them in trays seemed to work well; however, the transplanting process became the new bottleneck in the restoration process. In the words of Mike, "all we really figured out was how to murder lots of baby lupine." Newly planted seedlings barely lasted a week before shriveling to a crisp. It quickly became apparent to me that my role in this process was going to be to develop a more systematic approach to figuring out what these plants needed to germinate, grow, and most importantly, survive.

That year, I set to work gathering lots of data. By the end of the summer, the data helped us realize something that any beach goer could tell you. Sand can get very hot during the day. Walk out on the same stretch of beach at night and the sand will be as cold as the air around you. Throughout a twenty-four-hour period, the sand we were planting lupines in would regularly reach temperatures as high as 115°F to 130°F (46°C—54°C) during the day only to plummet to around 50°F to 60°F (10°C–15°C) at night. Experiencing such massive shifts in temperature each day can be very challenging for plants, especially in their first year of life. Thanks to the data, we had learned the most valuable lesson of all. If we wanted blue lupine to establish, we had to be a bit pickier about where we were putting our plants. By thoughtlessly planting them in bare, sandy soils, we were cooking them to death. It takes at least a year for blue lupine to put down a deep enough tap root to access water held deep in the soil, and here we were planting the tiny seedlings directly into the hottest, driest layers of soil. For them, it must have been like being locked in a parked car in full sun with the windows up and no water to drink.

From that point on, we were less focused on experimenting in the test plots and more focused on experimenting with the rest of the property. Each spring I would head out into the surrounding

A Rocky Start: How I Learned to Love Plants in the Bottom of a Quarry

grassland, tossing handfuls of lupine seeds as I went. I was curious to see if maybe nature would take its course and provide us with the solution to our lupine establishment issues. Obviously, these seeds would not be hand planted in the wild, so if there was any hope of getting lupine to grow at this site, it was going to happen wherever nature would allow. Following seed sowing, I would arrive on-site early each morning and go looking for signs of germination. At this point in time, I would have been happy to see even a single seedling poking up out of the sand. Though I wasn't collecting any proper data, I made some predictions about what I expected to find. My thought was that if I was going to find lupine seedlings, then I would find them in the larger spaces between other plants where competition would be minimal. After all, why would a plant perform better in the dense shade of a competitor?

Boy, I was wrong. When I finally stumbled across seedlings, most of them were found tucked in and among the shade of warm season grasses. There were some seedlings in the bare areas as well, but far fewer of them. Regardless, I was ecstatic to finally be seeing success in my lupine growing efforts. Had I had a smartphone at the time, I probably would have filled the entire memory with pictures of those little plants. My bosses were happy to hear about it as well. Perhaps hiring me wasn't such a bad idea after all? As that summer wore on, an interesting pattern started to develop among the lupine seedlings. Those that had germinated among the grasses were thriving, while the others were shriveling up and dying like the previous year's transplants.

It turns out that, for some plants, having close neighbors isn't always a bad thing. Under stressful conditions such as those found in the bottom of a sand and gravel pit, plants can often facilitate each other's growth, for a time, at least. In this situation, the clumps of warm season grasses were acting as nurse plants for the tiny lupine seedlings. Though far more common in dry, desert environments, the conditions in the bottom of that quarry were stressful enough that the situation appeared to be playing out in a similar way. What nurse plants do is create a favorable microclimate around their base where growing conditions are ever so slightly better compared to the surrounding landscape. They aren't doing this intentionally, of course. Plants, like all life, are not imbued with a sense of altruism. And, while this relationship may be beneficial at first, with the grasses providing some shade and humidity and blue lupine enriching the soil with an added boost of nitrogen, often these relationships unravel later in life. Those tiny seedlings being nursed along can eventually become fierce competitors for what few resources are available.

The very idea of plants interacting with one another, whether competing for resources like light, water, and nutrients, or facilitating one another during the toughest parts of the growing season, was completely foreign to me. Up until this point in my life, I had written plants off as sometimes pretty but more often boring. They might as well have been inanimate objects, for all I cared. School was no help in this, either. Plant-based units of our curriculum largely focused on the various steps involved in photosynthesis or memorizing the parts of a flower. At no point in time did anyone care to mention that plants were dynamic organisms struggling to survive, often in ways so completely alien to motile vertebrates like us. My experiences in the bottom of that sand and gravel pit were opening my eyes to a massive portion of the biosphere to which I had never given much attention. I was learning that plants had different needs and unique survival strategies. I was learning that plants interact with one another, often competing but occasionally cooperating. I was learning that plants were different from one another; what so often looked like a sea of sameness was really myriad distinct species I knew nothing about. Most importantly, I was learning that plants were absolutely fascinating and that, for someone with an obsessive mind, there were lifetimes of information waiting to be uncovered. It is hard to say how this particular situation between the grass and the lupine would play out in the long run. It would be many years before the lupines would grow big enough to be considered competitors. Either way, blue lupine was finally growing on its own at the site, and we were one tiny step closer to creating habitat suitable for the Karner blue butterfly.

Restoration is so often a slow process. Ecosystems don't form overnight. The vast majority of plants on our planet operate on timescales far slower than we can readily comprehend. It would be years before those first lupine seedlings would mature and start producing seeds of their own. It would be many years beyond that before the offspring of those first lupine would begin making flowers of their own. If Karner blue butterflies had any hope of being introduced to this site, we were going to have to wait for a thriving population of lupine and other plants to establish and function to some extent on their own. All restoration sites of this nature will need the help of humans into the future. We have disrupted natural systems to such an extent with invasive species, removal of life-giving fire, and fragmentation that little patches of nature like the one in the bottom of that sand and gravel pit need stewardship. This is OK, provided the will is there. Ecosystem stewardship builds deep relationships with nature. After all, look what it did to me. Still, my time working for the mining company was limited and when my contract was up, I had to move on.

By this point in time, the Karner blue butterfly had become a side note in my daily operations. Its reintroduction to the site was always the end goal, but plants had become the true objects of my obsession. I would spend hours in my office digging through scientific journals, reading blogs, and learning about various plant propagation techniques. I felt like I had been cheated of this enriching experience of getting to know plants, and it was time to play catchup. I realize that it is largely my fault for not even trying; however, I can't help but feel like most of society gives plants short shrift. Why were my introductions to plants focused on a small handful of points about the morphology (biological structure) and physiology? Why is it that we would spend days in our biology class talking about the behavior of animals like jellyfish or cheetahs and no time on the variety of ways in which plants interact with the world around them? Why do plants always get overshadowed by their pollinators?

Well, humans have serious biases. We like things we can relate to on some level and, to be fair, plants can be very hard to relate to. However, the crash course in botany I gained from working in that quarry showed me that plants can be both surprisingly relatable and incredibly alien all at once, and the combination of those two things is very exciting. I read science fiction to escape from the mundanity of human life. I delight in mentally exploring alien worlds and the bizarre lifeforms they support. But here on our planet was a group of organisms that felt just as different and exciting as any fictional alien species, and the best part was, they were tangible and real. I could meet these organisms in the flesh (cellulose?), get to know them, and even grow some of them myself.

I never got to see the restoration efforts in that sand and gravel pit through until the end. I found work on the other side of the country and had to move away. Though I was sad to leave my lupine work behind me, I could rest easy in knowing that I had made a contribution that would hopefully put the project on the pathway to success. Not only did I learn to love plants in that quarry, it also taught me important lessons in working with people, organizing projects, and needing to be adaptable in unpredictable situations. Working in nature is amazing, but rarely does it go as planned. When I think back on it, our struggle with lupines was probably one of the most instrumental periods in my life. Had we not had to overcome the transplanting bottleneck, I don't know if I would have put in hours of botanical research, and my obsessive mind may have never stumbled onto the fact that plants were incredibly interesting organisms worthy of respect and admiration. With plants situated firmly at the center of my focus, the world felt new and different, and I was so very excited to learn more. I had officially entered into a period in my life that I like to refer to as my green revolution.

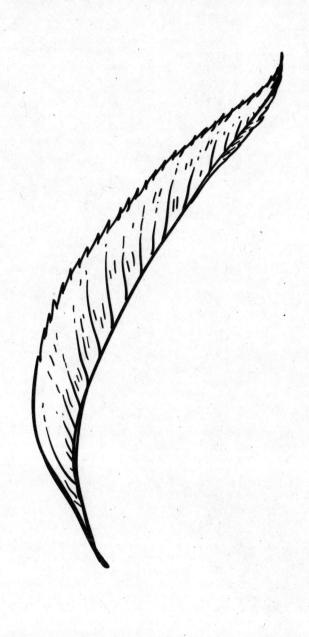

CHAPTER 2

My Own Green Revolution

My days at the quarry seem like a blur as so much has happened since then. My obsessive journey into the world of botany has turned me into a bit of a monster. Walk into my apartment today and you are greeted with a literal wall of tropical and desert vegetation. Go outside and you will find gardens packed with a diversity of native plant species. I have fallen into gardening in a big way. However, gardening is a bit challenging when you don't own land. Fortunately, while I was still living in western New York, I was lucky enough to be allowed some gardening space at my parents' house.

My parents live on about an acre of land forty-five minutes south of the city of Buffalo. The land was once forested, but the man who built the house cleared the lot of trees and allowed people to take topsoil from the back half of the property. The soil that remained is tough to work with. Thanks to New York's glacial past, most of my parents' property is comprised of heavy clay and unconsolidated rocky leftovers from the ice sheet's retreat some ten thousand years ago. Every swing of a shovel is met with a loud metallic clank as blade hits rock. The impact sends enough reverberation through the handle to weaken your grip. None of this mattered to me as all I wanted to do was grow plants. I was experiencing what I like to refer to as my own green revolution. Just like I got to know fish by figuring out how to keep them happy in small glass boxes, I knew I could learn more about plants by trying to grow them in the garden. And try, I did.

To garden is to fail...a lot. There is no way around the fact that you are going to kill many plants throughout your experience. You can't let this haunt you too much. Provided you aren't poaching rare plants from the wild or spending exorbitant amount of money on prized specimens, you can rest easy knowing that these failures eventually teach you how to succeed. Since those early days, my approach to gardening has been a bit different from many of my gardening pals. Beauty is and has always been secondary to me. First and foremost, I want plants that are interesting. Gardening for me is just as much of a chance to learn as picking up a book or leafing through a scientific journal, except that it is far more tangible. Each plant I put into the ground is like a newfound friend. I want to interact with plants from seed to flower and back to seed again. Most importantly, I want my gardening choices to be beneficial for the myriad other organisms living in and around my neighborhood.

Flip through the news today and stories of ecological destruction are commonplace. Insects, birds, and pretty much everything else is generally on the decline around the world. So much of this is due to the destruction of the habitats that support them. Habitat destruction comes in many shapes, sizes, and forms, but what most examples have in common are humans. Our need to conquer every last corner of the globe has come at great cost to Earth's biosphere. The most brutal and damaging forms of habitat destruction occur over huge swaths of land. Usually, these are done in the name of industrial progress of some kind. When we level thousands of hectares of a rainforest and replace it with palm oil plantations and cattle ranches, or when we blow off the tops of mountains to get at coal and other minerals locked away in the rocks, we are removing all of the habitat that land once supported. Such activities leave giant scars on the landscape that are impossible to ignore, but as tragic as these forms of destruction are, they all too often overshadow more insidious forms of habitat destruction happening all around us. Smaller forms of habitat destruction are often just as damaging, yet receive relatively little attention because humans have a hard time valuing small tracts of land as anything other than places waiting to be "improved" by development. Every new house and lawn that replaces a forest, every shopping center that replaces a wetland, every extra acre of corn field that plows over a tiny prairie remnant is destroying habitat as well. While larger forms of habitat destruction fragment ecosystems into tinier and tinier chunks, smaller forms of destruction make short work of what's left.

However, to simply say "habitat destruction" breezes over an extremely important point, and that is, plants *are* habitat. At some point in your life, you were undoubtedly introduced to the idea of a food chain or food web. These simplified diagrams describe the flow of energy (via food) through natural systems. Whether you picture this as a tiered triangle or a tangled web, there is no way around the fact that at the base or core of all of them sit plants. Via photosynthesis, plants utilize the energy given off by an unbelievably massive nuclear fusion reaction (our sun) taking place 92.147 million miles (146.5 million km) away in the vacuum of space to split water and CO_2 and make food. Without photosynthesis, Earth would be a completely closed system and close systems are, by definition, finite.

Thanks to plants, we have a constant source of new energy entering Earth's biosphere. Therefore, plants, or at least some form of photosynthetic organism (shout out to algae), sit at the base of all major food webs, with the notable exception of those deep-sea thermal vents at the bottom of our oceans. Plants produce the energy that is then dispersed throughout the rest of the living world. Just above plants on this chain or web sit arthropods, like insects. Just as the larvae of the Karner blue butterfly specialize on the leaves of blue lupine, most insect species on our planet are also specialists on specific groups of plants. Whether as larvae or adults, insects absolutely need the plants with which they share an evolutionary history to survive. When native plants disappear, so, too, do the insects they support. Insect decline may sound like a cool idea to those out there who are squeamish about insects, but it is very bad news for all forms of life, including us.

Though there are many different shapes, sizes, and lineages of herbivores on our planet, insects are arguably among the most important. Insects not only shape the abundance and distribution of plants via herbivory and other interactions that we will cover in later chapters, they are also food for so many other animals. Everything from fish, to shrews, to mice, to birds, to reptiles, and even some humans consider insects a regular and necessary part of their diet. In considering these simple facts, it doesn't take a leap in logic to realize that if we care about "charismatic" life forms like birds and fish, we need to care about insects. And, by the same reasoning, we absolutely must care about plants.

Thanks to my work in the quarry, I learned about the role of native plants in maintaining biodiversity. I wanted to do everything I could to cram as many native plants into my parents' landscape as possible. The act of planting native plants in their gardens felt like doing ecological restoration at home. By gardening with natives, one can so easily return some of the life that was lost when the lot was built in the first place, and that thought excited me to no end. My parents are lucky enough to have a decent chunk of forest at the end of their property, and since I was spending most of my working hours focusing on grassland and savanna species, I decided that I would give woodland gardening a try.

I have always had an affection for strange forms of life. I really like organisms that take a bit of effort to understand and love. This carries over heavily into my choice of garden plants. One evening, I found myself leafing through native seed catalogues on the lookout for something new and exciting. The first thing to catch my eye was its common name, "wild ginger." I knew enough about plants at this point to know that ginger in the culinary sense was derived from a tropical plant. There was no way I was going to be able to grow such a species in western New York, where winter temperatures regularly dip below -20°C. Whereas culinary ginger (*Zingiber officinale*) hails from the one hundred percent tropical family Zingiberaceae, the plant I was looking at was said to belong to the so-called "birthwort" family, Aristolochiaceae. I remember thinking to myself, "what the hell is a birthwort?" I had to know more about the natural history of this peculiar species.

Wild ginger goes by the scientific name *Asarum canadense* and sports two wonderfully green, heart-shaped leaves. The name wild ginger was given to the plant because its roots slightly resemble culinary ginger in taste. One must be careful about this, however, as wild ginger is well armed with a toxin called Aristolochic acid. Any herbivore, including humans, who wants to dine on this plant must contend with the carcinogenic and kidney-damaging effects of this compound. At least one herbivore has managed to turn this defense against the plant. Larvae of the pipevine swallowtail butterfly (*Battus philenor*) are not only immune to the toxic effects of these compounds, but by feeding on the leaves of this plant family, the caterpillars are able to sequester the toxins in their skin as a form of defense. Anything trying to eat a pipevine swallowtail caterpillar must also contend with the toxic effects of the plant the caterpillar was feeding on. Here, again, was another opportunity to provide valuable habitat for a butterfly by adding this wonderfully weird native plant to my garden. I hopped online and ordered some seeds that night. After I had confirmed my purchase, I needed to do more digging on this mysterious woodland plant.

The first thing I did was look up pictures of the flower. They were so unlike any flower I had encountered up to that point. Wild ginger flowers are borne at ground level under the dense shade of their heart-shaped leaves. Looking at them head on, each flower resembles a tiny hairy triangle with a concave pit in the center. Each hairy corner of the triangle is presented in an intriguing range of colors from nearly brown to burgundy or wine. Within that concave pit in the center were the reproductive bits, and those did not appear to resemble the stigma and anther anatomy I learned in school.

The bizarrely beautiful flower of wild ginger sits at ground level, below a canopy of heart-shaped leaves.

In fact, everything about their reproductive bits seemed counter to what I had come to expect from flowering plants. For instance, why would a plant produce its flowers hidden away under the shade of its leaves? Don't flowers need to be seen in order to be pollinated? Also, the flowering period for wild ginger is pretty early compared to the plants I was used to dealing with. Starting as early as the beginning of April in some places and lasting only until May, wild ginger likes to flower when most of the insects we consider pollinators are still waking up. It was like wild ginger didn't care if it got pollinated or not. Obviously, this couldn't be the case. How else would this species have stuck around long enough if it couldn't reproduce? A little more digging provided me with the second mind-blow of the evening.

Despite its unique look, very little is known about wild ginger pollination. There has been plenty of speculation for sure, but no one can seem to agree on who pollinates those strange flowers. Some have suggested ants; however, I find this to be an unlikely situation. The antimicrobial fluids with which ants coat themselves tend to render pollen infertile. Others have even suggested slugs crawling around on the moist forest floor are the pollinators. As they slime their way over the flowers, pollen may stick to their bodies and be transferred from flower to flower. This seems unlikely as well, especially considering that slugs are far more likely to eat flowers and foliage than provide much in the way of pollination services. By far the most widely adopted theory is that the best candidates for wild ginger pollination are small flies such as fungus gnats.

Fungus gnats are surprisingly active during the cold and damp weeks of early spring. As they emerge from their larval stages spent eating fungus in the soil, the adults begin looking for places to mate. Because they need to lay their eggs where fungus is abundant, the adults spend a lot of time in and around dark nooks and crannies on the forest floor. The thought is that these tiny gnats take refuge from nasty weather in the bowl of wild ginger flowers. The anthers of the wild ginger flower produce copious amounts of pollen that coats the gnats, which they then take to other flowers the next time they need to take shelter from the elements. There does not seem to be a lot of supporting evidence for this fungus gnat theory, either. To be honest, this isn't too surprising as pollination studies can be incredibly difficult. Even if you observe an insect visiting a flower, without proper experimentation and data, there is no telling how effective that visit will be for the plant. In fact, it is very likely that many wild ginger plants simply pollinate themselves.

Armed with this incredible natural history knowledge, I moved on to my second inquiry of that evening, what the hell is a birthwort? Another internet search told me that some of wild ginger's cousins have been used by humans to aid in the birthing process. Apparently, even the family name "Aristolochiaceae" is Latin for "the best for birth." Having never birthed a child myself, I cannot speak toward its efficacy. Regardless of its utility, the bizarrely alluring morphology and complex ecological relationships meant that this plant had to be included in my garden. Once the seeds arrived, I set to work planting them around the wooded areas of my parents' lot and the waiting game began.

I waited weeks for signs of life from those seeds. I would check on them daily but found nothing. As spring gave way to summer, my endless attention turned up nothing. I felt like I had failed at one of my first attempts to return a new native plant to the forest. It turns out this was probably no failure on my part. Many gardeners have noted that the seeds of woodland plants like wild ginger can take a very long time to germinate. This is one reason why native woodland plants aren't often found in nurseries; it's too hard to make any money growing and selling them. It is also one reason why woodland plants tend to suffer more from population decline than plants from more open habitats. It often takes their populations much longer to recover from disturbances. In the end, my seeds never amounted to anything. Luckily for me, my obsession with wild ginger was noted by a friend who kindly divided some of the wild ginger in their own garden and gifted it to me. With ginger in place, it was time to move on to other plants.

Later that summer I noticed a pattern in that wooded area. One half of the lot boasted a variety of wildflowers. There were small plants with delightfully speckled leaves and bright yellow flowers called trout lilies (*Erythronium americanum*), a handful of different violets including an interesting species with lavender colored flowers and a deep nectar spur called American dog violet (*Viola labradorica*), and even these little tufts of broad, waxy blue leaves belonging to what I thought were grasses but turned out to be a type of sedge with the uninspiring common name of broadleaf sedge (*Carex platyphylla*). However, on the other half of the lot, a single species seemed to dominate at the exclusion of pretty much everything else.

This bully of a plant would start out its first year as a dense clump of kidney-shaped leaves with scalloped edges. In its second year, the plant would throw up stem or two topped with a clump of white, four-petaled flowers. When you crushed any part of this plant, it gave off a smell reminiscent of broccoli and garlic. It didn't take much sleuthing to figure out its identity. What we had was an infestation of one of North America's most pernicious non-native plants, garlic mustard (*Alliaria petiolata*). If I was to better understand and deal with this invasive issue, I had to do my homework.

The first thing I learned about garlic mustard was that its introduction into North America was no accident. This mustard is native throughout a wide swath of Eurasia and has a long history with humans. Its ability to not only grow but thrive in disturbed habitats meant that garlic mustard likely followed human settlements wherever they went, and it didn't take long for people to figure out they could eat it. It makes a tasty addition to many a meal, provided it is harvested early. As is so often the case, once a plant becomes useful to our species, we like to take it everywhere we go. And take it, we did. Estimates place the introduction of garlic mustard into North America sometime in the 1860s.

With no predators and novel weapons, garlic mustard can
reach monoculture status here in North America.

The second thing I learned is that garlic mustard also engages in chemical warfare. It is what we call an "allelopathic plant," which means that it releases chemicals from its leaves and roots that inhibit the germination, growth, and reproduction of other plants around them. Garlic mustards' chemical cocktail is not lethal to the plants directly, but rather, it kills the mycorrhizal fungi that native plants rely on for improved access to water and nutrients in the soil. Since neither the plants nor the fungi have any evolutionary history in dealing with such compounds, the effect is stronger here in North America compared to garlic mustard's native range where species have had much more time to adapt. Understory plants like the violets and trout lilies I had become so enamored with are usually the first to suffer. Between overcrowding and the loss of their fungal symbionts, they can't hold their ground for very long. However, garlic mustard's impacts don't end with smaller species. There is even evidence that trees can fall victim to its chemical attacks as well. Trees also rely heavily on mycorrhizal fungi, and when garlic mustard populations explode, trees have a much harder time finding fungi to partner with and they, too, begin to decline. The negative effects of garlic mustard on trees has been found for a variety of groups including pines, maples, and even oaks. It just goes to show you that even size doesn't matter when novel weapons are introduced into an ecosystem that has never experienced them before.

Imagine if I had not caught the garlic mustard invasion for a few more years. Those plants would have continued to produce seed year after year, and those seeds would grow into more and more garlic mustard. Eventually, the fate of all the species in that tiny patch of forest would have been the same. As more native plants were edged out each year, so, too, would all the wonderful bees, moths, and other arthropods that use them for food and shelter. As these animals disappeared, all the birds that frequented the property would find it harder and harder to find food. The nuthatches, chickadees, and warblers would have to forage farther and farther away to be able to find enough food to raise their young. Through my battle with garlic mustard, I was making connections between the actions I was undertaking on this tiny plot of land and the effect on the greater ecosystem in which I lived. I was also learning that plants were not the peaceful little entities I once thought they were, and that only made them cooler in my mind.

There is no way around the fact that invasive species are second only to habitat destruction in pushing native species to the brink of extinction. However, the idea that we could completely eradicate all invasive species is a pipe dream. Even if we could, the measures employed to remove any species that has reached high densities on the landscape may do just as much harm as they do good. More often, the goal is to manage invaders enough that their numbers become controllable. Also, ecology being the massive grey area that it is, invasive species management becomes a case by case scenario, with effective methods differing from one region to the next. The point I am trying to make is that management of invasive species is a complex issue that can be tackled properly only if we all calm down, take a step back, and remember what our end goal should be—to protect and bolster native biodiversity.

As I was working toward increasing native plant diversity in my little woodland, I was also busy introducing more native plants to other parts of the yard. I didn't really have a plan other than to find as many different species as I could get my hands on. One summer afternoon, I was looking in the back corners of a local nursery when something tickled a few of my senses. I heard a gentle humming sound followed by some high-pitched squeaks. I looked around for the source and found two male ruby-throated hummingbirds engaged in an impressive aerial dogfight. Hummingbirds are incredible animals. They possess some of the most active metabolisms in the animal kingdom and must feed almost continuously to keep stoking their internal fires. They can't waste their time on low energy foods, which is why nectar makes up most of their diet. There is a ton of energy locked up in sugars, but it doesn't last long. This means that for hummingbirds to survive, they must have a nearly constant source of nectar nearby. Hummingbirds viciously defend good nectar sources because there is no telling when the next will come into bloom.

I was thinking to myself in that moment how interesting it was that most of the hummingbird encounters I have had were due to the artificial feeding stations we humans provide. As a child, I would stand under my grandpa's hummingbird feeder, gawking at the flashes of metallic green feathers as hummingbird after hummingbird vied for a spot at the feeder. I could think of only a few instances in my mind in which I witnessed a hummingbird feeding at an actual plant. There seemed to be a big disconnect here. Obviously, hummingbirds evolved long before hummingbird feeders were invented, so what did they eat when humans weren't around?

An answer came to me as soon as the hummingbirds sorted out their little spat. The victor chased off his rival and then proceeded to visit a small table full of something truly wonderful. I had never seen a plant like this before. Sitting atop tall, square stems were a series of small tubes that formed a dome. Out of the top of each tube exploded strange-looking flowers so red that I almost thought they were fake. The flowers didn't have the radial symmetry of a phlox or an aster. Instead, they resembled the gawking mouth of some sort of flamboyant tropical bird. Out of the top part of the flower extruded what looked like the reproductive parts. The bottom part of the flower looked like the fancy lip of an orchid blossom. As I watched, the hummingbird systematically probed each flower on the dome until it had drunk from all of them. It then moved on to the next stem and repeated the process. I suddenly felt the urge to attract more than insects to my gardens. I now wanted to focus some attention on providing hummingbirds with real food, not just the sugar water we put in plastic feeders. This plant seemed like a great place to start.

The stunning floral display of Oswego tea acts as a beacon for hummingbirds.

As soon as the hummingbird left the flowers, I rushed over to the table and grabbed a tag sticking out of one of the pots. The plant was called "Oswego tea" or, as it was known to science, *Monarda didyma*. The best part of it all was that it was a true native. *Monarda*s are a genus within the mint family (Lamiaceae) and most have adopted a similar blooming strategy with clusters of long, tubular flowers arranged at the top of the stem. Their gaudy inflorescences look like tiny firework displays frozen in mid-explosion. Many organisms besides hummingbirds will visit *Monarda* flowers, including a miniscule bee that goes by the clunky common name of the monarda dufourea (*Dufourea monardae*). It would be an easy bee to miss, maxing out around seven millimeters in length. To date, this tiny, unassuming black bee has been found feeding only on the flowers of *Monarda* species. Its entire lifecycle is tuned to the blooming period of these plants. I had learned all I needed to know at that point. I purchased a few pots of Oswego tea and upon returning home, found the perfect spot in the garden for them to grow.

The plants didn't do much that first year. Summer came and went, and all that I could see were those few scraggly stems. That all changed the following year. Each plant exploded in a profusion of new growth. By midsummer, they were in full reproductive mode. Each stem began to produce an inflorescence chock full of flower buds. Just below the buds sat a nearly radial set of smaller leaves that gradually turned from greed to deep red as the inflorescence matured. This color change reached a crescendo just as the first flowers erupted. Everything about their growth was now aimed at sex. Watching the process unfold, I felt like I was gaining deeper insights into the reproductive strategy of Oswego tea. The flowers didn't emerge all at once. Instead, flowers near the base of the inflorescence appeared first. Over the course of a week or two, new flowers emerged sequentially from the bottom up.

Within a day or two after those first flowers opened, my Oswego tea were already the stars of the garden. I witnessed multiple different bees attempting to take a drink of nectar. Most of the small bees seemed to be struggling. They couldn't reach far enough into the floral tube to access the nectar at the bottom. Some would give up after a few tries while others would switch gears and move to the upper tip of the flower to nab some pollen. Others were cleverer and had figured out that if they chewed a small hole into the base of the flower, they could siphon out some nectar with comparatively little effort. As exciting as all the bee activity was, something seemed off about their visits.

To be considered a successful pollinator, floral visitors must contact the reproductive structures of a flower in just the right way. If they fail to do this, pollination will not occur. That seemed to be the case with most of the bees I was observing. Those that circumvented the whole system by chewing holes at the base of the flower were nothing more than nectar thieves. Only the smaller bees that dangled from the tip of the flower as they collected pollen seemed to be doing the trick. Even so, growing such large and intricately showy flowers just to attract small bees seemed like a massive waste of energy for the plant. As I pondered these ideas in the garden, I thought back to the previous year when I was inspired to purchase them in the first place. Had hummingbirds discovered this new botanical bounty I had planted for them? They were certainly paying attention to the artificial feeder we had placed out on our back porch, so I knew hummingbirds were in the area. I just had to step back and be patient. I set up a lawn chair a yard or two back from the plants and waited.

Within less than an hour, something hummed over my head. I saw a flash of green descend onto the Oswego tea. A female ruby-throated hummingbird swooped in and went to work drinking nectar. Its grace stood in stark contrast to the clumsy feeding habits of the various native bees. As I watched, I realized that whereas bees were probably okay pollinators at best, without a doubt Oswego tea had evolved in response to the feeding habits of hummingbirds. The hummingbird would approach each flower rapidly but then slowly insert its long, slender bill down into the tube. As it drank the nectar at the bottom, its probing would bring both anthers and stigma into contact with its head. The process would repeat with each floral visit and the precision of it all was remarkable to say the least.

Indeed, research indicates that everything about the anatomy of Oswego tea flowers is primed to take advantage of hummingbird feeding habits and anatomy. The lower lip of each flower acts as a guide for the hummingbird's bill. Its shape ensures that the bird's head ends up in just the right spot for pollination to occur. I couldn't get over the complexities of the coevolutionary marvel that was playing out before my very eyes. Never before had I considered the importance of floral anatomy in the process of pollination. These flowers were no longer just an object of beauty or a chance to observe some native pollinators. My Oswego tea had become a portal into a much deeper appreciation and understanding of the fact that plants are simultaneously shaping and being shaped by life around them.

My gardening exploits did not end when I went inside for the day. My plant addiction had gone full throttle, and I had the need to surround myself with them at all times. This can be tough when you live in this temperate zone. In western New York, I had, at best, only about three months of the year to dedicate to outdoor gardening and plant exploration. When plants went dormant in the fall, so did a chunk of my happiness. If I was going to get through the long winters with my sanity intact, I needed to grow plants indoors.

Houseplants have always been a part of my life. Growing up, my mom dutifully tended and propagated a considerable pothos (genus *Epipremnum*) collection. Numerous pots and hanging baskets overflowing with these tropical vines were tucked into nearly every corner of the house. Every time my sister and I or even the dogs would brush by one and break a stem, my mom would go about rooting the stem back into the soil. She couldn't bring herself to throw away living things. Looking back on it, I think that her insistence that those severed pieces of stem were fully alive and worthy of another chance at growth must have instilled in me the idea that plants were living organisms, not just pretty decorations. This philosophy really stuck with me, whether I realized it or not, and greatly informed how I approached the houseplant hobby. Every new houseplant offered a chance to learn something new about that species and its ecology.

One day, my partner presented me with one of the coolest looking plants I had ever seen. Arising from a bright yellow pot were a handful of intensely succulent red stems, each tipped with a whirl of the most stunning leaves. Each leaf had a pleasing, velvety texture and was so deeply green in color that it appeared black in all but the brightest light. Bright red parallel veins ran the length of each leaf, interrupting the blackness like a glowing neon sign. Turn those beautiful leaves over and the bottom was colored in dark maroon. I was convinced I was looking at some new kind of succulent I had never encountered before. Imagine my surprise when she told me that this bizarre plant was an orchid.

It belonged to a group of orchids colloquially referred to as jewel orchids, no doubt due to the spectacular coloration of their stems and leaves. This particular species goes by the scientific name *Ludisia discolor*, and its very existence seemed like a contradiction to what little I knew about orchids at the time. First off, it was planted in regular potting soil. All the orchids I had encountered up to that point lived in pots full of chunky bark. This is because many of the orchids in the horticulture trade are epiphytes, living out their lives on the branches and trunks of trees where rich layers of soil can't form. Here in front of me was an orchid that wanted to live in soil. Another glaring contradiction was the disconnect between those thick, succulent stems and delicate, thin leaves. Where did this plant grow in the wild? The succulent stems suggested a desert, whereas those thin leaves suggested somewhere much wetter. Also, what was up with its intense color scheme? Finally, what did its flowers look like? I knew orchids produced some of the strangest looking flowers.

Surely, an orchid this bizarre must put on quite a show. Once again, my curiosity was thoroughly piqued.

As I searched the web for answers to my burning jewel orchid questions, I started to notice a pattern. Most of the information written about this plant consisted of growing instructions. These articles would start out by mentioning that jewel orchids were mainly grown for their decorative foliage and, though they flower with some regularity, the flowers themselves are nothing to write home about. Such floral disdain only made me want to see them even more. The other annoying thing about the articles I was finding was the fact that no one seemed to agree on where this plant originated. Some articles mentioned southern China while others suggested Thailand.

It is strange to me how disconnected even gardeners can be from the organisms they claim to cherish. For me, growing plants only deepens the respect I have for them as living things. I could not understand how someone could spend so much time with a plant and not have even the slightest curiosity about where it comes from and how it lives in the wild. At the very least, knowing such things about a plant will make taking care of it that much easier. Of course, not all gardeners are like this. However, my quest for knowledge on plants told me that a disturbing amount of "plant people" think this way. I needed to honor this strange orchid by learning as much as I could, and it took many more hours of searching before I could form something of a completely picture of this species.

The contrast between the white flowers and dark, velvety foliage of *Ludisia discolor* is absolutely breathtaking.

I was able to piece together that our jewel orchid is native to warm tropical forests throughout China and Southeast Asia. It does not grow as an epiphyte (an organism that lives on surface of a plant) but instead lives out its life in the dense shade of the forest floor. Most of the time, it barely roots itself into soil. Instead, it grows in the layer of partially decomposed plant material just above. Its succulent habit ensures that it always has water available when needed, and its dark coloration is thought to help it either make use of what little light is available under the canopy or to protect the sensitive photosynthetic machinery within its leaves from the occasional unfiltered sun flecks that dance over the forest floor as the sun traces its arc across the sky. Many understory plants share similar coloration, and yet, scientists can't fully agree on why. For now, I am happy with suggesting it serves multiple purposes.

It took an even longer time before I came across a more respectful take on jewel orchid flowers. Though smaller and less colorful than the flowers of the hybrid moth orchids (genus *Phalaenopsis*) you see at grocery stores, jewel orchid flowers are nonetheless fascinating structures. About once per year, mature specimens will begin to produce a fuzzy stem from the center of each whirl of leaves. The stem gradually elongates, eventually producing buds toward the tip. Each of those buds bursts open, revealing a flower that, at first glance, can best be described as a sunny side up egg. The sepals and petals are stark white while the column which houses the male and female organs is bright yellow. Juxtaposed against the darker tones of the rest of the plant, a cluster of blooming flower spikes is an incredible sight. How any plant enthusiast could denigrate this display is beyond me. Apparently, mass production of gaudy orchid hybrids has spoiled too many.

Jewel orchid flowers only become more fascinating the closer you look at them. Stare at one straight on and you will notice an odd quirk in its morphology. Whereas most orchid flowers are bilaterally symmetrical, meaning you can draw a vertical line down the middle and create two mirrored halves, the flowers of this particular species of jewel orchid are distinctly asymmetrical. The asymmetry is the result of a peculiar twist in the reproductive organs that always orients them to the right of the flower. This was very strange. How could such a peculiarity arise in arguably the most important organ of any flowering plant? The answer lies in how specific orchids can be regarding their pollinators.

Orchids differ from the classic model of flower pollination. Whereas we generally think of flowers wantonly dusting pollinators with prodigious amount of powdery pollen, orchids are much stingier. Instead of dusty anthers, orchids bundle up their pollen into paired sacks called pollinia. The pollinia sit atop a stalk that ends in a sticky structure called a viscidium. Bundling their pollen in this way means that each orchid flower really only gets one shot at a successful contribution of pollen. Having only a single chance to father new seeds may seem like a strategic downgrade, but consider the fact that orchids produce huge quantities of dust-like seeds. As such, each successful pollination event results in thousands upon thousands of potential offspring.

Orchids are also very specific about their pollinators. In most cases, only a single species of pollinator has what it takes to properly pollinate an orchid. Of course, there are exceptions to this, but usually those equate to a single genus of pollinator rather than a single species. The key to this specificity lies in the morphological complexity of the orchid flower itself. You can picture the pollination process as a lock and key scenario. The elaborate shapes, sizes, colors, and odors of orchid flowers have evolved over time so that only their specific pollinator can fit into the flower and properly pick up the pollinia. As the pollinator probes the flower, structures on the sepals and petals guide the visitor into just the right position to contact the pollinia. The sticky viscidium then glues itself onto the body of the pollinator in just the right spot so that when the process is repeated at another orchid flower, pollen is delivered precisely to where it needs to be to achieve pollination. There are so many variations on this theme that entire careers, journals, and books have been dedicated to teasing out the complexities of orchid sex. We will cover some unique examples in the next chapter, but let's get back to the lovely jewel orchid.

The asymmetry of the jewel orchid flower has everything to do with proper placement of the pollinia. Evidence from southern China indicates that butterflies are the target for *Ludisia* sex. And not just any butterfly will do. In the wild, our jewel orchid times its flowering period to coincide with the emergence of a species of butterfly with which many of us will be familiar, the cabbage white (*Pieris rapae*). Though today this butterfly enjoys a distribution that encompasses both northern and southern hemispheres, its original home range consisted of Europe, North Africa, and Asia.

When the cabbage white is not busy laying eggs on members of the mustard family, it is probing flowers for energy-rich nectar. Nectar-feeding insects like the cabbage white tend to look for plants that flower in quantity as abundance usually equates to more food. So, the jewel orchid can maximize its chances of getting the butterflies' attention by synchronizing its blooming period with the emergence of the cabbage white butterfly. The jewel orchid sweetens the deal by providing a tiny amount of nectar tucked away inside a miniscule pocket on the lower lip of the flower. When a butterfly visits a jewel orchid bloom, the asymmetry of the flower means that only a few spots offer a good enough foothold to allow the butterfly to feed. The specific placement of the footholds ensures that the pollinia will be attached in just the right spot. In this particular case, the right spot happens to be on one of the butterfly's legs.

Who knows why evolution favored the leg over other parts of the butterfly? Perhaps by sticking the pollinia to the leg rather than the proboscis or the head, the butterfly is less likely to spend time trying to remove it. It could also be that the orchid's morphology provided no other options and that placement on the leg simply works. Without knowing something about the ancestral relationships between these two species, speculation is all we have. What we can say for sure is that it works and, as the old saying goes, "if it ain't broke, don't fix it."

My time spent getting to know this incredible jewel orchid both horticulturally and ecologically was my first deep dive into the world of orchids. I was familiar with the common species offered in stores and nurseries, and I had encountered one or two species in the wild, but I never gave them much thought beyond that. In learning about the natural history of this species, I came to the realization that orchids were going to require further study on my part. If this single species had evolved such an incredibly intricate life history and the orchid family represented the greatest diversity among flowering plants, then there were seemingly endless amounts of information to pack into my mind. Overnight, orchids had gone from obscure curiosities into a full-blown obsession. I had contracted orchidelirium, a disease that has afflicted countless individuals since the Victorian era. Lucky for me, orchids are a readily available part of the gardening trade. I could find ethically sourced orchids and bring them into my home for more in-depth appreciation and study, and that is exactly what I did. Over the years, my orchid collection has grown immensely, and I have enjoyed getting to know each and every one of them.

By growing plants in and around my home, I was gaining a deeper respect for the fact that species matter. Plants like our jewel orchid are not just numbers on a biodiversity chart. Along with all the other species that share its habitat, they represent important snapshots in evolutionary history that connect the past and future in ways we are still trying to understand. Gardening was also connecting me with the environment in a much bigger way. By doing what I could to mimic the environmental needs of that jewel orchid, I was beginning to understand just how delicate its habitat could be. It wouldn't take much habitat disturbance to kill off swaths of this wonderful orchid, and remember, it is but one player in its forest ecosystem.

The connection my plants were giving me to the rest of the natural world was staggering. My green revolution was both a blessing and a curse. Learning to love a species only to find out how dire its plight in the wild has become is sobering. More often, though, successfully growing a species is extremely rewarding. Not only does it help me escape from the trials and tribulations of everyday life, it also inspires me to do better for plants and the planet. After all, one cannot wallow in despair forever. If we care, we must do everything we can to fight for species we love. One of the ways I do this is by getting to know plants so that I can relate their incredible stories to the rest of the world. Every year that goes by, I watch my plants grow, mature, and flower, noting how slight changes to my routine or their placement on a shelf alters their trajectories. I watch as new leaves unfurl and they outgrow their pots. I gain insights into their biology all along the way. Flowering is easily one of the most rewarding experiences of growing any angiosperm. It is (usually) the plant's way of telling you that it is getting what it wants. I say "usually" because some plants can and will flower themselves to death. Fortunately, this does not happen to me very often. Still, as beautiful as flowers can be, I am amazed at how little we understand the reproductive habits of the plants with which we share our planet. I was lucky to find pollination research that focused on our jewel orchid, but for so many other plants, such information doesn't exist. There are so many questions waiting for answers.

Growing up, I always got the sense that most of the "low hanging fruit" of science had already been plucked. It is easy to get lost in the minutia of more recent scientific investigations, and they often make you feel like there is nothing left out there to discover. Not so! Mysteries abound in nature, especially when it comes to more "obscure" forms of life such as tiny orchids that grow deep in some remote tropical rainforest. Mysteries aren't reserved for faraway places, either. Plenty of mysteries exist in our own backyards due to the simple fact that no one has thought to ask or taken the time to investigate.

Taking the time to look is a key aspect of doing any form of science, but it is especially true for anyone interested in studying pollination. Because of my podcast, I have had the great fortune of sitting down to talk with some incredible pollination ecologists. The one thing that unites their work, regardless of what group of plants or pollinators they study, is just how difficult pollination studies can be. As I mentioned, just because something visits the reproductive organs of a plant, that doesn't mean it is an effective pollinator. Also, as the jewel orchid taught me, pollination is not always as simple as bee visits flower, gets covered in pollen, and repeats the process over and over. Pollination runs the spectrum from general to extremely specialized, with all different shades in between. Also, it turns out that pollination isn't just for flowering plants. A handful of gymnosperms and even some bryophytes have gotten into the animal pollination game, and they go about it in some surprising ways. If my time growing orchids and other curious plants has taught me anything, it's that plant sex is strange. That is exactly what we will be talking about in the next chapter.

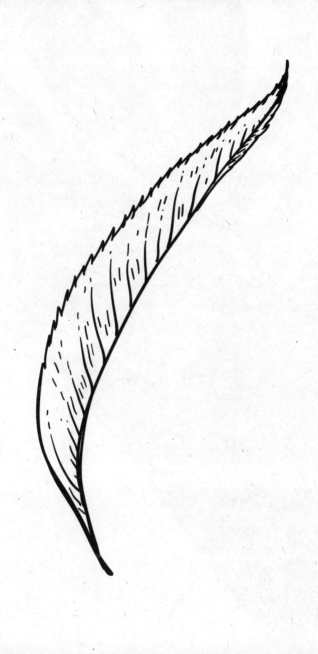

CHAPTER 3

The Wild World of Plant Sex

I magine that to successfully reproduce, you first had to create little bundles of cells from underneath your arms, each equipped with only half of your genetic material. Once those packets had matured, you then had to catapult them around you with little care as to where they ended up. Most of your bundles never make it. Some end up in trees, others end up in ponds, and still others get blown way out into the ocean. No matter, there will be more. Eventually, one of your bundles lands in just the right spot. A few other, unrelated bundles settled there too. Now, imagine that bundle "germinates," and from it begins to grow something that looks absolutely nothing like you in any way, shape, or form.

What grows from that bundle is a wrinkly, fleshy sheet that is orders of magnitude smaller than you. As the sheet matures, strange structures begin to form. Each of those structures resembles a mini volcano turned upside down. At the same time, the other bundles are going through the same process. They grow until all the fleshy sheets in this spot are nearly touching one another. The sheets don't do much but sit there. They are waiting for something.

Then, one day, it starts to rain. Eventually, a tiny pool of water has formed under the sheets, triggering a change. The volcano-like structures begin to ooze their contents into the water. With the help of powerful magnification, one would observe even tinier cells swimming, yes, swimming, through the water column. Each of those swimming cells is a sperm, complete with flailing flagella that power their movements. A few sperm eventually lock on to a chemical signal. They start homing in on the source, swimming as fast as they can in that direction. The signal is coming from some of those volcano-like structures on the other sheets. The sperm swim up inside and are greeted by an egg. Sperm and egg then fuse, combine their genetic material, and become a zygote. The resulting zygote now has the full complement of DNA needed to build an organism. And build, it does. Over time, it becomes obvious that the developing zygote is not growing into another fleshy sheet. Instead, it is beginning to resemble you! It doesn't take long for the zygote to develop into a fully functioning organism of its own. It quickly outgrows the fleshy sheet, whose job is now done. Reproduction complete, the sheet eventually withers away, leaving only your growing child behind. At this point, your reproductive efforts have been a success, and the process can repeat itself again.

It would be very weird if we reproduced like this, wouldn't it? As alien as this process may sound, reproductive cycles like this evolved on our planet hundreds of millions of years ago and continue to function in much the same way today. The scenario outlined above is that of fern reproduction, and the process is referred to as an alternation of generations. What we recognize as ferns with their beautiful feathery fronds represent just half of the fern lifecycle called the sporophyte generation. Its main function is to produce the spores. The spores are dispersed into the environment. When they find a suitable place to germinate, they grow into the other half of the life cycle known as the gametophyte generation. Gametophytes are not only free living, they are what produce the sex cells (sperm and eggs) that will eventually unite to from another sporophyte.

Ferns are vascular plants that can trace their lineages back to the Devonian Period, some 360 million years ago, but they are by no means the oldest lineage of land plants. Things get even weirder when we look at nonvascular plants like mosses, liverworts, and hornworts. Their gametophyte stage is the dominant stage of the lifecycle. When you look at a patch of moss attached to a rock or a colony of liverworts growing along a streambank, you are looking at gametophytes. Look closely at these colonies and you will see that many are adorned with different structures. In hornworts, these resemble horns (hence the name). In liverworts, some resemble stalked capsules while others look like tiny umbrellas. In mosses, the capsules are frequently borne on long stalks. Each of these structures represents the sporophyte stage, which contains two set of chromosomes, thus making it genetically distinct from the rest of the plant. However, unlike ferns whose sporophyte generation lives independent of the gametophyte, the sporophytes of hornworts, liverworts, and mosses are completely dependent on the gametophyte for water and nutrients. Most aren't even capable of photosynthesizing. The relationship between sporophyte and gametophyte almost resembles that of a parasite and host. Of course, because they are an essential phase in the reproductive process, the relationship is both necessary and beneficial. Still, it is a very odd arrangement, especially when contrasted against animal reproduction.

Even sex among flowering plants is strange when you think about it. Imagine needing a third party to reproduce, and that third party is largely unaware that it is getting involved the sex lives of other organisms. That third party can come in many forms. The means by which plants achieve pollination is a subject in and of itself. Throughout this chapter, we will be looking at some incredible examples of plant sex but always keep in mind that there are many more out there. Plants have evolved seemingly endless ways by which to get the job done.

It took learning about the sex lives of mosses for me to realize that they are just as complex and interesting as other more familiar plant species. I used to use words like "lowly," "diminutive," and "primitive" to describe them, but even the slightest bit of familiarity with these plants will demonstrate that such words have no place in the bryophyte lexicon. Mosses and company are tried and true evolutionary successes that have stood the test of time. They exhibit remarkably complex lifestyles. Their means of ensuring their genes make it into the next generation will blow your mind. Some have even evolved reproductive relationships we thought were reserved for flowering plants.

We are all too familiar with how insects like bees enable pollination. A bee visits a flower, drinks some nectar, collects pollen and then repeats the process at another flower. Well, this scenario plays out in a similar way in some moss species, only the "pollinators" are not moving pollen, but sperm instead. One of the best studied examples of this involves a moss you undoubtedly have growing in your neighborhood. It is one of the hardiest mosses on our planet and it goes by the badass common name of fire moss.

Fire moss, or *Ceratodon purpureus* as its known to science, is tough as nails. It thrives in disturbed habitats across the globe and can handle every sort of environmental onslaught we throw at it. From fires to soils contaminated with toxic heavy metals, there is no stopping the indefatigable fire moss. However, its ability to not only tolerate but thrive under extreme environmental stress isn't the only interesting aspect of its biology. Fire moss gets even more intriguing when you observe its sexual habits. Whereas moss species living in wet environments can let water do the work of uniting male and female sex cells, fire moss doesn't always have that option. Instead, fire moss has evolved a means of actively recruiting tiny arthropods called springtails to transport sperm from male to female moss.

Springtails are fascinating little creatures. They are small and hard to see, but springtails can be found anywhere there is enough moisture to keep them from drying out. As such, humid clumps of moss offer springtails a perfect place to live. Fire moss colonies provide ample habitat for springtails. In a biological equivalent of paying rent, fire moss coaxes the springtails into doing their sexual bidding. Both male and female fire moss stems emit a series of complex volatile scent compounds whose chemical composition makes them particularly attractive to springtails. The tiny arthropods will crawl all over the stems, looking for the source of the chemical lure, and that is exactly what fire moss needs. Like pollen on the hairs of a bee, fire moss sperm sticks to springtails. After a short visit to a male stem, sperm-coated springtails set off in search of richer scent trails. They follow them *en masse* until they end up on female stems. Female fire moss stems have been shown to emit more chemicals than male stems, which means springtails spend even more time exploring them and are thus more likely to brush off sperm on the female sex organs. Essentially, you can think of springtails as fire moss "pollinators."

No one is sure what it is exactly about the fire moss scents that attracts springtails. Perhaps they mimic springtail pheromones or maybe some sort of food reward. Either way, the relationship appears to work nicely for both organisms. Fire moss gets to reproduce while the springtails find a nice humid place to hang out and forage for food. It is also interesting to note that the ancestors of both mosses and springtails appeared long before the evolution of flowers. Springtails can trace their origins back to the Devonian, some 400 million years ago, and what we recognize as mosses have been around since at least the Permian, some 298 million years ago. Personally, I think it is highly unlikely that the fire moss-springtail relationship is unique and may even indicate that pollination-like relationships may be as old as terrestrial plants themselves.

I am amazed by how many extant plant lineages still produce swimming sperm. Swimming sperm is a throwback to the early days of plant evolution. Because they arose from aquatic algae, a sperm's ability to swim to an egg helped increase the chances of reproduction. However, swimming sperm is not a trait that ended with spore-producing plants. Other lineages carried this trait along with them throughout their evolution, only they became less dependent on water in the process. Plants did this by bundling up their sperm into tiny, protein-rich packets called pollen. Pollen protects the sperm from desiccation, allowing it to be carried much greater distances without harm. Cycads are one group of plants still alive today that exhibit this reproductive biology.

The Wild World of Plant Sex

Cycads do not get the attention that they deserve. The most famous cycad today is probably the poorly named Sago palm, *Cycas revoluta*. Palms, they are not. Cycads are gymnosperms, meaning they do not enclose their seeds within an ovary. By extension, this means that they do not produce flowers or true fruits. The reproductive structures of cycads consist of cone-like structures called strobili. Individual cycads are either male or female, never both, and males produce strobili chock full of pollen. Because cycads evolved long before familiar insects like bees, wasps, and butterflies came onto the scene, many have written off cycad pollination as being the product of wind alone. This is certainly true for some species, but in recent years, scientists have been able to demonstrate that insects are an essential factor in the pollination of many cycads. What's more, these relationships often appear to be extremely specific.

By far, the bulk of cycad pollination services are being performed by beetles. This is not surprising because, like cycads, beetles evolved long before bees, butterflies, or birds. In some cases, beetles utilize cycad cones as places to mate and lay eggs. For instance, male and female cones of the South African cycad *Encephalartos friderici-guilielmi* are popular mating spot for at least two types of beetle. Beetles will visit male cones, mate, lay eggs, and their larvae feed on the pollen as they develop. Once mature, adult beetles emerge from the male cone, carrying pollen all over their little bodies. Inevitably, some of these pollen-laden beetles wind up visiting female cycad cones and the process repeats itself. As the beetles go about their business, they successfully pollinate the female.

Beetles also share the cycad pollination spotlight with another surprising group of insects—thrips. Like springtails, thrips, too, belong to an ancient order of insects. Generally speaking, thrips are considered plant pests. However, for Australian cycads in the genus *Macrozamia*, they are important pollinators. Thrips prefer dark places that are both cool and dry to feed and breed, and male *Macrozamia* cones provide just that. However, if the thrips were to remain in the male cones only, pollination wouldn't occur but luckily, male *Macrozamia* have evolved a thrip-evicting trick. Via a special metabolic process, male *Macrozamia* can generate heat within their cones, which drives up humidity levels in the process. At the same time, the cones also start to produce compounds called monoterpenes. The combination of heat, humidity, and monoterpenes aggravates thrips living within, driving them out in search of a new home. Inevitably, some of these pollen-covered thrips wind up crawling inside female *Macrozamia* cones.

Female *Macrozamia* cones do not produce heat or volatile compounds, and thrips find the environment within much more to their liking. As they settle in, the thrips bring pollen into contact with the ovule. Once this happens, the pollen grain fuses with the ovule and begins to grow a tube. At this point, the pollen acts almost like a parasite, sucking up nutrients from the ovule tissue and destroying it in the process. When the tube finally makes contact with the egg, the pollen grain bursts, releasing the sperm inside. Cycad sperm look nothing like human sperm. They resemble little seeds covered in concentric rings of beating flagella. It's those flagella that are needed at this point in the game. Each sperm is like a tiny submarine, capable of swimming around inside the ovule until it locates the female gametophyte. Then, and only then, is fertilization accomplished. In this way, cycads no longer require water from the outside world for sex. Instead, each female creates its own watery environment within the ovule, ensuring that sex can happen on their terms.

As you can see, many of the plants all too frequently referred to as "primitive" go through extremely complex processes to achieve sexual reproduction. They are just doing so on a scale that is often difficult to observe, and they're partnering with some unexpected organisms in the process. Only when you eventually get to the subject of pollination in flowering plants will you begin to recognize some familiar players like bees. There is no denying that flowing plant pollination is one of the most important process on Earth, especially in the context of the food we eat. In fact, we hear so much about bees and flowers that we all too often take insect pollination for granted. However, many of the tactics flowing plants employ to entice floral visitors are just as wonderfully alien as those we just learned about.

Aside from bees, butterflies are often conjured up to express our appreciation for the process of pollination. Images of colorful wings flitting around equally colorful flowers are ingrained our psyche. However, for all the attention they receive, butterflies are surprisingly poor pollinators. Most species lack the anatomy needed to properly pollinate most flowers. That is not to say butterflies don't do any pollination whatsoever. As we have already learned, at least one species of jewel orchid absolutely requires butterflies to successfully reproduce. The jewel orchid isn't alone, either. Many of the larger flowered lilies have also coopted butterflies for sex.

Species like the Turk's cap lily (*Lilium superbum*), named for the idea that its flowers somehow resemble traditional Turkish hats, have evolved to utilize the wings of large butterflies to successfully transfer pollen from anther to stigma. In fact, the Turk's cap lily makes this process very easy to observe in person thanks to the dark coloration of its pollen. Find a nice sized population of blooming lilies and take a seat. It won't be long before swallowtail butterflies descend upon their gaudy blooms for a sip of nectar. The downward facing flowers force the butterflies to alight upside down to drink from the ample nectaries at the base of each petal. As they feed, the butterflies constantly adjust their position to keep their footing, flapping their wings as they fight to maintain balance. Each flap of their beautiful wings brings them into contact with the coffee-colored pollen on each anther. Eventually, large brown patches of pollen develop on the outside of each wing. These patches make it painfully obvious who has been visiting lilies that day. As these butterflies move from flower to flower, their large wings inevitably contact the sticky stigma at the tip of the long style and pollination is achieved.

As interesting as butterfly pollination examples can be, they don't hold a candle to the moths. Moths are the real rock stars of the lepidopteran pollination game. Not only are moths far more numerous than butterflies, they are much more effective at pollination, too. A perfect example of this involves a relationship so specific and strange that it may make your skin crawl.

Brown patches of lily pollen can be seen on the wings of this pipevine swallowtail.

Consider the sex lives of leafflower trees (genus *Glochidion*). Native throughout parts of Asia, these trees have struck up an obligate pollination mutualism with a small family of moths. Their flowers have become so tailored to the behavior of these moths that no other insect can pollinate them. In return for their services, female moths are provided with an edible place to lay their eggs—the fruit of the tree. If that wasn't strange enough, one species of leafflower tree takes this relationship to a whole new level. *Glochidion lanceolarium* of southeastern China literally holds its pollinators captive.

Unlike the vast majority of pollination which occurs by happenstance as the pollinator searches out a reward of some sort, pollination of the leafflower tree by its moth is deliberate. The flowers offer no other rewards besides the yet-to-be developed seeds tucked inside the ovaries. Gravid female leafflower moths locate *G. lanceolarium* blooms with the help of a special perfume that the tree has tailored specifically to this species of moth. When they find a tree in bloom, female moths first visit the male flowers where they pick up some pollen. They then move on to the female flowers where they deposit the pollen into a special chamber that can only be accessed by the moths' proboscis. Once pollen has been deposited, the female leafflower moth then locates the ovaries of the flower, and using a needle-like ovipositor, deposits eggs within the undeveloped fruits so that their larvae hatch right next to their food source.

It doesn't pay for leafflower moth larvae to be gluttons. They can only eat one or two of the dozens of seeds developing within the fruit, otherwise, the tree will abort that fruit, killing any larvae within. For most leafflower tree/moth pairings, once the larvae have eaten their allotted number of seeds, they will then chew an escape hole in the fruit, pupate, and emerge as fully-formed moths, ready to repeat the process. Such is not the case for *G. lanceolarium* and its moth. Instead of chewing their way out, the larvae of the leafflower moth are held captive within the fruit for nearly a year. Yes, you read that right, a year. Cut open a fruit of this leafflower tree and there is a chance you might find a fully-formed moth waiting patiently inside one of the swollen ovary chambers. Only once the fruits have matured and begin to split open will the moths within be released. This just so happens to occur right as the new crop of flowers are opening, and thus, the process repeats itself year after year. By holding the moths captive, the tree is literally controlling when its one and only pollinator is available to do its reproductive bidding.

Insects are among the most effective pollinators available to plants, but what happens when they aren't around? There are places on this planet where conditions are too tough to support large insect populations. This is often the case high up in the alpine zones of mountains. Weather can be so brutal at high elevation that pollinating insects are often hard to come by. Under such circumstances, evolution has devised alternative means for pollination. One of my favorite examples can be found in South America. Along the craggy peaks of the Andes, from Chile into Argentina, grows a strange alpine plant commonly referred to as Darwin's slipper (*Calceolaria uniflora*).

The flowers of Darwin's slipper look more alien than botanical.

When you take a step back and look at Darwin's slipper, you can see that it conforms to what one can expect out of a plant adapted to high-elevation habitats with its compact mounds of leaves borne on short stems. We refer to plants that have adopted compact growth of this nature as "cushion plants." It is an adaptation against blustering winds and cold temperatures. Compact growth minimizes exposure to the elements and provides the interior of the plant with a more favorable microclimate that protects sensitive tissues. The flowers of Darwin's slipper must not have gotten the compact growth memo, though. Each bloom is surprisingly large, sticking well up above the plant for all the world to see. They are also outlandishly colorful and oddly shaped. Their base color is a yellowish orange, but starting near the top and increasing in density toward the bottom are flecks and blobs of bright red pigment. Attached to the bottom lip of the flower sits a fleshy patch of tissue the color of porcelain. The whole floral package is quite a spectacle. Such outlandish-looking flowers must obviously attract something, but as already mentioned, pollinating insects are in short supply up in these mountains. What are present are birds, specifically a species of seedsnipe, and they are exactly what Darwin's slipper requires to have sex.

Seedsnipes are adorable little birds that like to hang out in high-elevation grasslands throughout much of South America. They exist on a plant-based diet and spend most of their time holding territories and grazing on the seeds and fruits of alpine plants. Spend some time perusing patches of Darwin's slipper growing in proximity to seedsnipe territories and you will notice that their flowers receive high levels of damage, specifically on the lower lip where the white appendage is located. In fact, when seedsnipes are around, that white appendage often disappears entirely. Further observation would reveal that seedsnipes regularly visit patches of Darwin's slipper and proceed to peck off and eat those white appendages. As a seedsnipe pecks away at the bottom of the flower, the anthers and stigma bash against the bird's head. Pollen is dusted onto the seedsnipe's head and any pollen that was already there from a previous visit is simultaneously dabbed onto the stigma. Thus, Darwin's slipper achieves pollination. But what's in it for the birds? Why are they so interested in that fleshy, white floral appendage? As it turns out, those white appendages are high in sugars, making them an irresistible and nutritious part of the seedsnipe diet.

The relationship between Darwin's slipper and the seedsnipe may seem pretty outlandish, but it certainly works for both species. The lack of insects in its alpine habitat has driven Darwin's slipper toward an alternative pollinator, and a unique one, at that. However, insects and birds are not the only flying animals to enter the pollination game. Bats, too, are important pollinators in many ecosystems. Recent focus on pollination as an ecosystem service has brought a handful of bat-pollinated plants into the limelight. Articles celebrating the role of bats in banana or chocolate farming abound, but these flying mammals pollinate so much more than domesticated crops. For instance, many cacti rely heavily on bats for their reproductive needs.

The mighty saguaro cactus (*Carnegiea gigantea*) is one such species. It can take over thirty years for a saguaro cactus to reach maturity and begin producing flowers. From that point on, it wastes no time. Bundles of cream-colored flowers are produced at the apex of the cactus, and it is thought that the growth of those charismatic saguaro arms is largely a way for mature individuals to increase their reproductive potential. Saguaro flowers are small compared to the size of the cactus, but what they lack in size, they make up for in abundance. Dense clusters of saguaro flowers begin to open in the evening and produce most of their alluring scent throughout the night. Though a wide variety of animals will visit saguaro blooms, the main pollinators are the nocturnal lesser long-nosed bats (*Leptonycteris yerbabuenae*). Interestingly, it has been found that saguaro flowers produce a suite of amino acids that help female bats sustain lactation while raising their young, making saguaro nectar an invaluable food source for these flying mammals. Both species are icons of the desert southwest of North America and both could not exist without each other.

Bat pollination is fascinating and can get pretty weird the more you investigate. Take, for instance, the case of a beautifully strange member of the pea family called the sea bean (*Mucuna holtonii*). You can find the sea bean and its relatives growing throughout the tropical forests of Central and South America where they dangle their incredible inflorescence from tall branches like a beautiful floral chandelier. Like the saguaro, the bulk of sea bean pollination is done by bats. As you can probably imagine, finding food in a dense tropical rainforest is a complex task for nectar-feeding bats. For many years, it was thought that such bats relied solely on scent rather than sonar to find their food. After all, unlike insects that move around, flowers are stationary and don't lend themselves well to echolocation. However, such assumptions were abandoned once we developed microphones sensitive enough to pick up bat calls.

The mighty Saguaro is an icon of the American Southwest.

With the help of high-powered microphones, scientists discovered that nectar-feeding bats utilize unique frequencies for echolocation, which are primed for resolving finer details than those of their insect feeding cousins. Whereas many bat-pollinated plants position their flowers in easy to reach places for bats, the sea bean takes things a step further. Its pendulous inflorescence is ringed in flowers that, although highly derived for a pea, are nonetheless representative of the family. All the key pieces are there—the wing-like petals, the keel, and a single large petal known as the banner—and it's that banner petal that is the sea bean's secret to successfully attracting bats.

The banner petal acts as a nectar guide, though not in a strictly visual sense. Sea bean flowers open at night when nectar-feeding bats are foraging. The flowers begin to emit a scent, which lets the bats know roughly when and where a meal is available. Once a bat has homed in on the inflorescence, that banner petal comes into play. Supremely adapted to the specific frequency of these nectar-feeding bats, the banner petal of each virgin flower reflects sound waves over a greater range of directions than the clutter of the surrounding forest, thus helping the bats zero in on the exact location of their next meal. Visiting bats cannot access the nectar by hovering as we see in other bat pollution systems. Instead, bats must land on the flower to feed. It's the weight of the bat that each flower requires to complete the process.

The pendulous inflorescence of the sea bean hangs
like a chandelier from the canopy above.

When a bat lands on the flower, it triggers an explosive mechanism that snaps the anthers outward, causing pollen to erupt from the flower and spattering it all over the bat's back. No harm is done to the bat in the process, and it will continue to feed until the nectar is gone. Once the bat drinks its fill, nectar production ceases. However, the flowers don't senesce at this point. Flowers remain in good condition long after being pollinated. Yet, bats somehow know that an already-visited flower is not worth the effort of a visit and subsequently avoid those individual blooms. But how do they know?

The answer again lies in that banner petal. Once a flower has been triggered by a bat, its shape changes. This alters the way in which bat sonar is reflected. Bats soon learn that flowers with this altered shape no longer offer a nectar reward. The plant benefits from this because it reduces the chance that a bat will end up depositing its pollen right back on the flower it came from. Thus, cross pollination is encouraged, and the plant is still able to maintain a large floral display that continues to attract bats to flowers that have yet to be pollinated. It's win-win when you think about it. Bats maximize their meals, and the plant maximizes its chances of cross pollination.

The sea bean isn't alone in tapping into bat sonar, either. Back in 2011, yet another sonar-related relationship between a plant and a bat was discovered in Cuba. The plant in question is known only as *Marcgravia evenia*. When mature, *M. evenia* dangles an inflorescence from the tip of one of its branches. Like the sea bean, the *M. evenia* inflorescence resembles a tiny chandelier adorned by some of the strangest flowers in the botanical kingdom. The flowers themselves point downward, presenting a spray of reproductive organs that face the ground. Just below the flowers hangs another ring of what looks like bright red urns. These urns fill with nectar, drawing in visitors from far and wide in search of a drink. However, most of the animals that visit *M. evenia* flowers, which includes everything from bees to moths and even hummingbirds, are not large enough to contact the reproductive organs as they feed. They are, in effect, nectar thieves. Only Pallas's long-tongued bat (*Glossophaga soricina*) is the right size to get the job done and evolution has equipped *M. evenia* with exactly what is needed to get the bats' attention.

Just above the inflorescence sits a single dish-shaped leaf. It may not look like much compared to the complexity of the floral display below but that single leaf is the key to successful pollination for *M. evenia*. That leaf functions like a tiny reflector for the bat's sonar. When the sonar of Pallas's long-tongued bat hits that leaf, the echoes that return to the bat are not only strong and multidirectional, they are also recognizable and invariant, meaning that they provide a reliable and unchanging signal to the bat each and every time. Once the bat has locked onto the inflorescence it swoops in and begins to feed. As it laps up nectar from the cups below, its head contacts the reproductive organs of each flower, picking up and depositing pollen in the process.

I don't want to paint the picture that bats are the only mammals to have gotten involved in plant pollination. Far from being occasional evolutionary oddities, many plants have coopted small furry critters for their reproductive needs. A great example of this can be found in a species of hedgehog lily known as *Massonia depressa*. This species is native to arid regions of South Africa, and each winter, two broad leaves are produced that lay flat on the ground. Between those leaves emerges a wonderfully spiny inflorescence (hence the name hedgehog lily) made up of numerous rigid, cream-colored flowers. The flowers emit a yeasty odor, which attracts rodents like gerbils.

The hedgehog lily rewards its rodent visitors with extremely viscous nectar. And when I say viscous, I mean it. Its nectar is four hundred times more viscous than other nectar solutions with a similar sugar content. The viscosity makes it much easier for the rodents to effectively lap up the nectar. It also causes a lot of pollen to stick to their face in the process. When not eating, rodents spend a lot of time grooming, so the plant must ensure that some pollen stays on the fur long enough to make it to the flowers of another hedgehog lily.

Arthropods, birds, and mammals: as odd as some of these examples are, they represent the bulk of pollination services on our planet. As such, they have received most of the scientific and public attention. However, there is another group of animals out there that we are only really starting to appreciate for the surprising number of pollination services they provide—lizards. Recent decades have seen an uptick in lizard pollination studies, and they suggest that we have overlooked reptiles for far too long.

Lizards may seem like strange pollinators. They don't have feathers or fur for pollen to stick to, and most of the time we picture lizards dining on meatier fare like insects. To be fair, pollination services are not widespread among reptiles, but are instead restricted to a few different groups like geckos. Also, lizard pollination tends to be more common on islands than on any mainland. This is not surprising, as islands are metaphorical playgrounds for evolution. The isolated nature of islands means that it can be a bit of a lottery in terms of what kinds of flora and fauna survive the journey over water or air and arrive on an island in any number to establish a functioning population. As such, islands are frequently full of "empty niches" ripe for the taking, and both lizards and plants are quite good at doing just that.

One such example of a plant relying exclusively on lizards for pollination can be found on the island of Mauritius. When not in bloom, *Nescodon mauritianus* is a rather unassuming plant with its lanky stems and whirls of lance-shaped leaves. During the flowering season, however, it would be a hard plant to miss. Surprisingly large bell-shaped flowers dressed in the most wonderful shade of lavender are borne from the axils of the leaves. The stems that support them are not terribly strong, which means each flower tends to droop toward the ground. The most remarkable feature of these flowers is not their size, shape, or positioning. It's the nectar they produce. Peer up into a flower and you will see that it is oozing with bright red nectar. Since it was first described, the striking contrast of the red nectar against the lavender corolla has had botanists wondering what exactly these plants are attracting.

The bright red nectar of *Nescogon mauritianus* contrasts nicely with
its lavender flower. Photo Credit: Dr. Martin Christenhusz

The obvious answer seemed to be birds. It is no mystery that birds see in color much in the same way as we do. However, multiple experiments and many hours of observation have since revealed that birds are not the best pollinators for these flowers. In fact, they function mainly as thieves, stealing nectar without actually contacting any of the necessary floral parts for pollination to occur. Luckily for the plant, birds aren't the only animals paying a visit to the flowers.

Living alongside *N. mauritianus* are a few species of brightly colored day geckos in the genus *Phelsuma*. Like *N. mauritianus*, they, too, are endemic to Mauritius and can frequently be found visiting the flowers of many different plants in search of a sugary meal. Experiments have revealed that these geckos are particularly attracted to shades of red and yellow, which means the red nectar of *N. mauritianus* really gets their attention. It is possible that because these geckos rely on brightly colored markings when interacting with each other, the selection for colored nectar has been favored in the evolution of gecko-pollinated plants like *N. mauritianus*. More work needs to be done to say for certain. What we can say for sure is that their feeding habits once inside the flower puts the geckos in direct contact with the reproductive parts. Thus, these adorably colorful day geckos function as the most effective pollinators for this remarkable and rare plant.

I think more than any other symbiosis in the natural world, pollination is viewed as an altruistic venture. There are seemingly endless poems, essays, and more devoted to waxing poetic about how pollination represents the harmony inherent in natural systems. As an ecologist, I find such views to be nice, but misguided. We want to look at nature and see something pure and good. Certainly, I find nature to be innocent, but it is by no means good or bad. Petty human emotions like these do not bog down natural systems. Everything in nature comes down to surviving long enough to get your genes into the next generation, and pollination is one of the most interesting ways any organism has evolved to do just that.

The more you dive into the science behind our understanding of plant pollination, the more you begin to see it for what it really is: an epic tug of war between organisms, each trying to gain as much from the interaction as possible while at the same time giving as little in return. It is very important to remember that, from beginning to end, reproduction takes a lot of energy. Cones, spores, sepals, petals, pollen, nectar, seeds, all of these take a lot of resources for plants to produce and maintain. Considering the cost of reproduction from an evolutionary standpoint, you can begin to see why it may be advantageous to cheat the system a bit. Up to this point, we have only considered pollination systems that provide some form of reward to the pollinator, but there exist many plants on our planet that offer no rewards at all. Such plants rely on deception to trick pollinators into visiting their flowers. There are many fascinating ways in which a plant can pull off such a ruse.

The brush-like protuberances on the lip petal of the grass pink
orchid (*Calopogon tuberosus*) are a form of food mimicry.

Some of the most studied examples of pollinator deception involve food deception. Such a scenario only works if the flowers produce structures that resemble food or if the floral display is presented in such a way that the pollinator can't assess whether food is present or not without first paying a visit. It should come as no surprise at this point that orchids are masters at both types of food deception. One of my favorites can be found in a group of terrestrial orchids called the grass pinks (genus *Calopogon*). The grass pinks comprise about five distinct species scattered throughout eastern North America as well as Cuba and the Bahamas. On the lips of their flowers, the grass pinks produce a series of yellow, brushy protuberances that resemble clusters of pollen-laden anthers. These are not anthers because, as you may remember, all orchid pollen is bundled up into sticky sacs called pollinia. Instead, those brushy protuberances are merely mimicking pollen, tricking bees into thinking they are only one visit away from a protein-rich meal. When a bee swoops in and lands on the lip, its weight causes the petal to bend at the base like a lever, slamming the bee into the actual reproductive organs. Stunned, the bee flies off with no reward and pollinia stuck to its back. If the orchid is lucky, the bee will be fooled at least one more time by an unrelated grass pink orchid.

Another intriguing example of food deception in the orchid family can be found in the elder-flowered orchid (*Dactylorhiza sambucina*) of Europe. This species presents its flowers in such a way that passing insects can't tell whether food may be present or not, forcing anyone hungry enough to pay its flowers a visit before they find out they've been duped. How the orchid does this is absolutely fascinating. The inflorescence of the elder-flowered orchid is made up of a dense cluster of flowers that come in two distinct color morphs, purple and yellow. Plants of the two morphs look so drastically different that one could be excused for thinking they were two different species. Even more strange is the fact that in any given population, you can find flowers of both colors in close proximity to one another. To understand why this is, we must consider who pollinates the elder-flowered orchid.

Bumblebees are no dummies. Far from being mindless drones whose sole purpose is to benefit the colony, these industrious insects are quite capable of learning and memory. For deceptive plants that rely on bumblebees for pollination, the bumblebees' penchant for remembering presents a particular challenge. Thanks to their highly tuned search image, bumblebees can quickly learn which plants are worth visiting and which plants are not. Any plant that doesn't give them what they want will quickly be shunned. This is where having different colored flowers comes in handy.

Scientists have discovered that the color ratios of any given elder-flowered orchid population are under what is referred to as "negative frequency-dependent selection." This is one of those fancy pieces of jargon that scientists invent for new theories, but it is pretty simple to understand. Here's how it works: naïve bumblebees that visit a non-rewarding flower of one color (say purple, for this example) soon learn that purple flowers aren't worth their time. Those bumblebees are then much more likely to visit a flower of a different color (yellow). It just so happens that in this case, the plant with yellow flowers turns out to be the same species of orchid. Each bumblebee will then visit a few yellow flowers before it figures out that they, too, offer no reward. By this time, however, the orchid has already succeeded in tricking the bee into transferring its pollen to an unrelated individual, and thus, pollination is achieved without wasting energy on any form of reward system.

This dual colored flower system results in some interesting floral dynamics among elder-flowered orchids. Tricking the bees with flower color means that in any given orchid population, plants with the rarer flower color receive more visits. Because flower color is under genetic control, in this case, the rarest color morph will gradually rise in frequency within that population. Once that color becomes the dominant flower color, the reverse happens, and the first color is then visited more often. Over time, this causes back and forth shifts in flower color that eventually settles on some sort of stable ratio of purple to yellow flowers. Thus, anyone botanizing a high-elevation meadow in Europe can find purple and yellow flowered orchids living together. By tapping into the bees' natural foraging tendencies, this non-rewarding orchid species can reproduce without having to invest valuable energy into floral rewards.

Orchids aren't alone in utilizing food deception for pollination. Other plant families have evolved trickery as well but, in my opinion, they are mere palate cleansers compared to the next example. Some of the most remarkable examples of pollination via food deception can be found in a strange group of vining succulents hailing from the milkweed family (Apocynaceae). The genus *Ceropegia* is something of a taxonomic mess. There are many described species ranging over parts of South Africa, southern Asia, and Australia, and botanists are working hard at trying to figure out how many represent true species, how many represent nothing more than unique morphs of different species, and whether or not they should continue to exist in the genus *Ceropegia* at all. All of that is neither here nor there. What you need to know about these plants is that they produce incredibly complex flowers that rival the intrigue and beauty of any orchid.

The flowers of most *Ceropegia* look like something you would see in an experimental art gallery. Many consist of tube-like structures outlandishly colored and presented in all manner of sizes and shapes. Often, the opening at the top of the tube is adorned by an equally extravagant hood that allows entry through holes along its side. The actual reproductive structures are what one would expect out of the milkweed family, but they are tucked down in at the base of the floral tube. No nectar is produced and, like orchids, *Ceropegia* pack their pollen into pollinia.

Though not readily apparent, *Ceropegia* flowers function like tiny pitcher plants. Insects that come to investigate lose their footing and fall into the floral tube. The inside of the tube is either too slippery for insects to crawl back out or it is lined with downward pointing hairs that prevent upward mobility. Of course, the flowers aren't trying to eat their captives but instead force them into picking up and depositing pollen. As fascinating as it is that these plants trap their pollinators, the most mind-blowing aspect of their pollination strategy is how their pollinators are attracted in the first place. To demonstrate how incredible *Ceropegia* pollination can be, let's look at a species that is gaining popularity among houseplant growers looking for something different.

The plant in question is sometimes called the parachute plant (*Ceropegia sandersonii*) thanks to the brightly colored, parachute-like hood that covers each flower. It's a vining species native to South Africa that produces long, fluted white flowers decorated in green splotches. The holes at the top of the hood are lined with delicate hairs that blow this way and that with the slightest bit of air movement. The parachute plant is very specific when it comes to pollination, requiring a unique group of kleptoparasitic flies to get the job done. Kleptoparasites are any species that make their living by stealing food from other organisms, and the flies that pollinate the parachute plant just so happen to specialize in sucking the juices out of bees that are in the process of being eaten by spiders. When a spider captures a bee, the bee releases stress compounds that are carried on the wind. For kleptoparasitic flies, these stress compounds function like a dinner bell. As the spider liquefies the hapless bee, the flies sneak in for a drink. Because their diet is so specialized, scientists were surprised to see so many of these kleptoparasitic flies visiting parachute plant flowers, so they decided to take a closer look.

Their analyses revealed something so bizarre that even the most seasoned science fiction writer would have trouble topping it. The scent compounds released by parachute plant flowers are surprisingly similar to the stress compounds released by dying bees. In fact, roughly sixty percent were an exact match. By releasing compounds that make each flower smell like a dying bee, the parachute plant lures in hungry kleptoparasitic flies. They fly into the opening at the top in search of a dying bee and fall down inside. As they scramble around the interior of the flower looking for a way out, they inevitably pick up and hopefully deposit packets of parachute plant pollen. After about a day of imprisonment, the flowers wilt, releasing the flies inside to hopefully make the same mistakes again. If the parachute plant example wasn't incredible enough, consider the fact that each species of *Ceropegia* seems to have its own unique strategy for attracting its own unique set of pollinators. *Ceropegia crassifolia*, for example, is pollinated by fruit flies, and its flowers emit compounds that mimic yeast fermentation. As more scientific investigations place *Ceropegia* at the center of their focus, a whole world of complex scent mimicry is surely to be revealed.

We can all empathize, to some degree, with food deception. I, too, could be tricked into doing things with the promise of a tasty meal. It works because all organisms need to eat, but it's not the only form of deception plants use to get pollinated. There are plants out there that tap into an urge even more primal than the need to eat. Such plants utilize the promise of sex to lure in pollinators. I'm not talking about offering pollinators a place to mate or lay eggs. When I say sexual deception, I mean tricking a pollinator into thinking it's having sex with a female that is actually a flower. As far as anyone can tell, the intended target is always a male. Hang out in a bar on the weekends and it doesn't take much imagination to understand why. Males of most species will go to great lengths to try and get their genes into the next generation. As such, it doesn't take much to trick a male into thinking he has a chance. Sexually deceptive pollination syndromes are so effective that they have evolved multiple times independently among flowering plants, but the undeniable champions of the sexual deception game are, you guessed it, the orchids.

The fuzzy edges and iridescent patch that decorate the flowers of the mirror bee orchid (*Ophrys speculum*) are thought to mimic a female bee.

If you live in Europe, then you don't have to travel far to find some classic examples of sexually deceptive flowers. The bee orchids (genus *Ophrys*) provide us with a wonderful window into this bizarre world and they do so in stunningly beautiful ways. The flowers of each species offer a mosaic of colors and texture that have captured the imagination of scientists and plant lovers throughout the centuries. Even Charles Darwin couldn't help but bask in the glory of these plants. He dedicated a considerable amount of time to trying to understand them better.

Bee orchids grow native throughout much of Europe, North Africa, the Canary Islands, and parts of the Middle East. Phylogenetically speaking, they are yet another confusing group of plants. Estimates of the number of species range from as few as twenty to as many as 130. All the confusion stems in large part from the fact that before the advent of affordable molecular techniques, the main way in which species were defined was via the morphology and color of their flowers. Bee orchids certainly don't lend well to the morphological species concept as their flowers can vary a lot from one population to the next. Yet, as confusing as bee orchid floral variety may be to us, it has everything to do with the evolution of this genus.

These amazing plants earned the name "bee orchids" because that is exactly who they recruit for pollination. And what a relationship it is. The bee orchids have evolved a means of tricking male bees into thinking their flowers are receptive female bees looking for love. The key to the ruse lies in the lip of their flowers. Each is adorned with bumps, tubercles, and hairs, which look and feel a lot like the abdomen of a female bee. An impressive array of color patterns on the lip further sell this mimicry. The flowers of some species even exhibit iridescent patches along the center of the lip that resemble a pair of bee wings. The ruse does not end with visual and tactile cues either. Far more convincing than their appearance are their floral odors.

Bee orchid flowers emit chemical compounds called "allomones," which closely mimic the pheromones released by receptive female bees. As you can imagine, the result of this chemical mimicry means that the relationship between bee orchids and their pollinators is often extremely specific. Each species of bee orchid produces allomones tailored to the type of bee they are trying to attract. For some, this means attracting males of only a single species. For others, the allomones attract a small handful of different bees with a genus. Regardless of the specificity, for a time, male bees find these flowers irresistible. They swoop in and aggressively attempt to mate with the flower in a process referred to as pseudocopulation. The floral morphology really comes into play at this point in the game as the male must be positioned in precisely the right way to effectively pick up and deposit the pollinia. All those hairs, color patterns, and protuberances are thought to aid in the positioning. The sexual ruse presented by bee orchids is so convincing that often male bees will preferentially visit bee orchid flowers over actual living, breathing female bees. However, the ruse doesn't last forever. Eventually, males learn that they are wasting a lot of energy on a farce and will start to avoid bee orchid flowers. This doesn't seem to be much of an issue for the bee orchids as a single plant can produce tens of thousands of seeds from a single pseudocopulatory event.

There is a pattern among sexually deceptive orchids in that most seem to target bees or wasps for their trickery. The key to understanding this pattern lies in the reproductive dynamics of these insects. Because male bees and wasps live for only a short time and really only function as sperm donors, the cost of distracting males each season is often negligible to bee and wasp populations as a whole. Thus, the overall cost of the ruse is not enough to drive pollinator evolution in the direction of males being more selective about mate choice. The variety of bee orchids stand as a testament to this fact, but on the continent of Australia, there exists another group of orchids that really put the reproductive habits of hymenopterans to the test. These plants have taken pseudocopulation to a whole new level by actively messing with the sex ratios of their pollinators.

"Tongue orchid" may not be the most appealing common name for an otherwise fascinating group of plants, but members of the genus *Cryptostylis* certainly make up for it by producing strange and complex floral displays. Most of their sepals and petals are highly reduced, often resembling little wisps of thread radiating out from the center of the bloom. The same cannot be said for the highly modified lip petal, which has been folded into complex shapes, furnished with tiny maroon hairs, nobly bumps, and dressed in shades of red, yellow, and orange. As beautiful and strange as these flowers are, one would be hard pressed to figure out the identity of their intended target. Unlike the bee orchids, tongue orchid flowers don't seem to resemble anything at all. Indeed, the tongue orchids are living proof that sex-crazed male insects don't require a precise replica of a female to be fooled.

The flowers of the large tongue orchid (*Cryptostylis erecta*) may not
look like a female wasp, but they certainly smell like one.

Instead of bees, the tongue orchids of Australia target male wasps in the genus *Lissopimpla* by mimicking their sex pheromones. They are very effective at it, too. As in the bee orchid example, male wasps find tongue orchid flowers irresistible and will often prefer to "mate" with them over actual females. However, tongue orchids appear to have gotten a little carried away with their trickery. Unlike many other reported cases of pseudocopulatory pollination in which males cease their mating attempts as soon as they figure out that the object of their affection is a flower and not a female, there is something about tongue orchid flowers that coaxes the male wasps to mate to completion, depositing their sperm onto the flower. Male wasps are not much different than male bees in that they are essentially little more than sperm donors. They don't live long after mating, and evolution has armed them with only so much sperm to go around. Therefore, you would think that wasting their precious few mating attempts on a flower would be disastrous for wasp populations. However, this is where tricking bees and wasps comes in handy.

The key to the stability of this pollination syndrome lies in Hymenopteran DNA. Queen wasps and bees are genetically haplodiploid. I will spare you the details on that, but basically what it means for Hymenoptera is that females are produced when sperm fertilizes an egg while males are produced via unfertilized eggs. Any queen whose eggs go unfertilized that year will produce the next crop of males while the rest of the fertilized queens produce the next crop of females. The tongue orchids have, unknowingly of course, tapped into this system to their benefit. When tongue orchids are in bloom, male wasps in that region waste a lot of sperm on their failed attempts to mate with flowers. This means that there are fewer males available to fertilize female wasps, causing the females in that population to produce lots of male offspring.

Tongue orchids are effectively skewing the sex ratio of their pollinators in such a way as to benefit their pollination needs. By causing female wasps to produce more males, the orchids are ensuring that there will be more naïve males in the population the next time they are in bloom. It also means that there will be even fewer female wasps in the population. With fewer females to mate with, male wasps become even less choosy and thus more likely to "mate" with orchids. As damaging to the wasps as this may sound, it actually works out to their benefit, too. With more sexually frustrated males flying around, each female wasp has a greater chance of being fertilized. Every year, a delicate balance is set between orchid and wasp. Because of the unique mating system that has evolved in Hymenoptera, the orchids have been able to maintain this pollination strategy with little harm to their pollinators.

As I said in the beginning of this chapter, plant sex is strange. The deeper you dive into this world, the weirder things get. Pollination is so much more interesting than the simple "bee visits a flower" model, and it is a shame that most of the examples of plant sex being taught only ever scratch the surface of this subject. With so many bizarre examples of plant sex on our planet, we shouldn't be starved for discussion topics. In fact, as I quickly realized while writing this, limiting oneself to only a handful of unique cases is a far more difficult task. Still, I hope that the little taste I have given you here has stoked a fire for learning that leads you to dive even deeper into the wild world of plant sex. For the sake of progress, however, it is time to move on. Once plants have done the deed of making new offspring, they must then disperse them out into the environment. Spores and seeds must find themselves new territory so that they can germinate, grow, and repeat the process again. In the next chapter, we will look at the various ways in which plants move around the landscape.

CHAPTER 4

Plants on the Move

The idea of plants moving around the landscape may seem silly at first. After all, the one thing plants can't do is walk. However, consider this fact: a mere 11,000 years ago, most of the northern hemisphere was buried under an ice sheet that, in places, was over a mile thick. The Earth underneath was scraped clean of all but the smallest forms of life. Yet today, these once glaciated regions are full of plants. Forests exist well into the Arctic Circle and even in places where trees can't grow, plants like mosses, herbs, and grasses take their place. If plants couldn't move, then how did they recolonize these regions? The answer to this conundrum lies not in some secret ability to uproot themselves and stumble around the land like a triffid. Instead, it lies in propagules like spores and seeds. Plants don't need to walk to conquer new territory because they travel over generations. How they do this varies from plant to plant and is one of the most interesting aspects of plant biology.

One of the simplest means of getting seeds out into the environment is to utilize the wind. Wind is especially useful to spore producers like bryophytes (hornworts, liverworts, and mosses) and ferns. Spores are extremely tiny structures, and it doesn't take much of a breeze to get them airborne. In fact, under the right conditions, spores can be carried many thousands of miles from the plant that produced them. The reality of this is made apparent on oceanic islands whose flora skew heavily in favor of spore-producing plants. Take the example of the Moss God of Hawai'i. A single male spore of a species of moss called *Sphagnum palustre* landed on Kohala over 20,000 years ago and has been cloning itself ever since. Hawai'i is one of the most isolated regions on our planet and the only way life could arrive there (before humans got involved) is by wind or by sea. That single spore made it there by crossing great expanses of the Pacific Ocean on air currents and it certainly isn't the only plant to do so. But spores don't have the wind all to themselves. Myriad seed producing plants have evolved a means of utilizing wind to disperse their seeds.

Like spores, wind dispersed seeds are often small, but that isn't always the case. Large seeds can utilize the wind; they just require special anatomical features to get them aloft. As a kid, I used to delight myself for hours by playing with maple seeds. We would gather them up by the bucketful and try to find the tallest place we could access to throw them off. I was a little less daunted by heights in those early days, so it usually meant climbing a tree. Once in position, we would grab a handful of seeds and toss them into the air, watching with joy as the thin blade caught the wind, sending each into a helicopter spin. Even then, it amazed me how far the spinning motion could carry these seeds. On windy days, the effect was even greater. By evolving an aerodynamic blade that keeps the seeds aloft, plants like maples can beef up their seeds with more energy reserves, giving the embryo within a head start on life once it germinates.

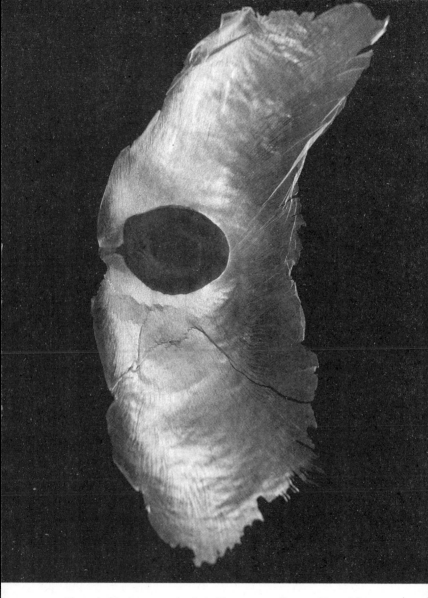

The seed of the Javan cucumber looks like some sort of botanical hang glider.

In my humble opinion, the master of wind dispersal is not a moss, nor is it a maple. The title must certainly belong to the Javan cucumber (*Alsomitra macrocarpa*), a unique relative of the garden cucumbers we eat each summer. At home in the tropical forests of the Malay Archipelago and the Indonesian islands, this vining plant takes wind dispersal to a whole new level by producing seeds that look like hang gliders. The Javan cucumber is a vine that winds its way up into the towering canopy of the trees that support it. Like its relatives, it produces a gourd or melon-like fruit the size of a football called a pepo. However, this pepo is not filled with juicy flesh. Instead, it is packed full of coin-shaped seeds sporting thin, membranous wings. Upon ripening, a hole opens in the bottom of the pepo and the seeds gradually fall out. This is when the spectacle begins.

With a wingspan of nearly five inches (thirteen cm.), the seeds of the Javan cucumber are immediately airborne. Like a giant morpho butterfly, the seed glides away from the plant bobbing up and down as it falls. The wings are situated in such a way that each dip produces lift. As the seed loses speed on the upturn, gravity overcomes it and pulls it back toward the ground. However, as the falling seed gains more speed, the wings catch the air again, sending it aloft. Depending on the weather conditions at that time, this process can continue for great distances, carrying the seed far away from its parent. In fact, sailors in the region have reported finding seeds of the Javan cucumber on the deck of their ship far out at sea. With such an effective form of seed dispersal, it is strange we don't find the Javan cucumber growing over a wider range.

Other groups of plants take advantage of the wind in different ways. Who reading this right now has never seen images of a tumbleweed blowing across the desert? Indeed, few images are more iconic of the American West than the tumbling of a tumbleweed. It is worth noting that the most common species of tumbleweed, *Salsola tragus*, is not native to North America at all. It's from Russia. Also, it's not a thistle but rather a member of the Amaranth family (Amaranthaceae). Regardless of its provenance and familial ties, the tumbling motion of this species is all about seed dispersal. When the seeds mature, the plant dries up. As water leaves its tissues, the branches curl inward, forming the whole plant into a ball. Eventually, a weak point at the base of the stem breaks, freeing the plant from its anchor. All dried and crispy, the lightweight husk of the plant is free to blow around the landscape. As it tumbles, the seeds are scattered in every direction. The fact that this invader can be found all over North America is a testament to how effective the tumbleweed strategy can be.

As convenient as the wind can be for dispersal, it does have its drawbacks. Wind can be unreliable. It isn't always present when propagules are ripe. Also, the strength and direction can be very idiosyncratic, and there is no guarantee that a propagule carried on the wind is going to end up in a favorable spot. Indeed, the vast majority of spores and seeds carried by the wind are doomed to failure. They can be blown too far and end up in a whole new habitat or they could simply wind up in the ocean or some other body of water. Dispersal distance is a real issue to consider. Plants need to get their propagules far enough away from their place of origin because their offspring will eventually compete for the same resources. However, if they travel too far, there is a good chance they will not find a good place to germinate. I suspect that the evolution of other forms of dispersal were driven in large part by issues such as these. For some plants, the solution to such problems is downright explosive.

Ballistic dispersal just sounds fun. I always picture ballistic dispersal as a crouched plant dressed in military garb, sighting a mortar loaded with its seeds. As silly as this imagery may be, it isn't too far off the mark for some plants. Take, for instance, a strange group of ferns in the genus *Angiopteris*. Not only can these ferns grow to massive sizes (they are among the largest ferns on our planet), but they utilize a unique form of spore dispersal not found in other ferns. Along the leaflets or pinnae of their giant fronds are tiny spore-bearing structures called sporangia. Each sporangium contains a multitude of tiny, spherical spores sitting in liquid. As the spores ripen, the sporangia start to dehydrate. The walls of the sporangia start to buckle in on themselves, creating pressure within the interior of the structure. The pressure differential eventually becomes so great that any liquid remaining between the walls of the sporangia instantly vaporizes, leading to a process called cavitation. Cavitation ruptures the sporangia with such intensity that the spores contain within are blasted out into the environment at great speed. They may not travel very far compared to their wind-borne cousins, but the explosive power does the job of getting the spores away from their parent but still within the bounds of welcoming habitat.

Some flowering plants utilize explosive power as well. One of my favorite examples can be found ambling over the ground throughout the Mediterranean regions of western Europe, northern Africa, and parts of temperate Asia. The squirting cucumber (*Ecballium elaterium*) truly earns its name. It seems especially fond of growing along roadsides and other highly disturbed areas, so if you find yourself traveling through one of these regions, you stand a good chance of crossing paths. After flowering, the fruits gradually swell into modestly sized, fuzzy version of the sorts you expect from the gourd family. From the outside, the fruits are fairly unassuming, but their gentle outward appearance belies a more turbulent story within.

As the fruit reaches maturity, the tissues surrounding the seeds begin to break down. The breakdown of this material creates a lot of mucilaginous liquid, causing internal pressure to build. And I mean a lot of pressure. Measurements have revealed that at peak ripening, pressures within the fruit can reach upwards of twenty-seven times the amount of atmospheric pressure we experience when standing at sea level! At the same time, the attachment point of the stem or "peduncle" weakens. With all that pressure building, it isn't long before something must give. This is exactly the moment when the squirting cucumber earns its name. The stem breaks away from the fruit, creating a small hole. Within a fraction of a second, all the pressurized mucilage comes rocketing outward, carrying the precious cargo of seeds with it. The result is remarkable. Seeds are launched anywhere from six to twenty feet (one to six m.) away from the parent plant. It is no wonder then that this is an incredibly successful species throughout its range.

The squirting cucumber looks very unassuming for
a plant with such explosive potential.

A similar explosive seed dispersal mechanism can also be found in the lodgepole pine dwarf mistletoe (*Arceuthobium americanum*). It makes its living as a parasite on, you guessed it, the branches of lodgepole pines (*Pinus contorta*). This species is peculiar, even by mistletoe standards. It does not photosynthesize and thus produce no leaves or chlorophyll. As a result, the entire body of the plant takes on a yellow to orange hue that looks more fungal than botanical. Because it absolutely requires a lodgepole pine on which to grow, the dwarf mistletoe cannot afford to leave seed dispersal up to chance. But it doesn't produce fleshy fruits like many of its cousins. Instead, it opts for ballistics.

The lodgepole pine dwarf mistletoe is one of a handful of plants that is capable of producing heat. Thermogenesis as it is called, is an interesting phenomenon in plants that often involves metabolic processes similar to that of animals like hummingbirds. Whereas most cases of thermogenesis in plants is thought to aid in pollination and seed development, the lodgepole pine dwarf mistletoe produces heat as a means of dispersing its seeds. As the fruits reach maturity, the mitochondria within their tissues kick into high gear. Mitochondria, you might remember, are the so-called powerhouses of cells. All the energy they produce creates heat. All that excess heat leads to a buildup in pressure within the dwarf mistletoe fruit until, BOOM, the fruit explodes, ejecting the seeds out into the surrounding canopy.

The force of the explosion is so great that dwarf mistletoe seeds can reach speeds upwards of sixty m.p.h. (a hundred km.h.)! This is an incredible feat for such a small plant. I can only imagine what it must feel like to get hit by one of these seeds. Picture yourself climbing a stately lodgepole pine only to be mowed down by mistletoe marksmen. The current record for dwarf mistletoe seed dispersal is sixty-five feet (twenty m.). Explosive seed dispersal is a real boon for a plant that needs to live on trees. If seeds were to just tumble away, the likelihood of them landing on a stem or branch of another tree would be small. By shooting them out into a canopy filled with tangles of branches, the seeds are way more likely to hit their target. It also helps that dwarf mistletoe seeds are sticky. Once they hit a branch, the seeds stick in place and can get underway with germination.

Not all cases of explosive dispersal require such complex chemistry. For some plants, all that is needed is for their fruits to dry. One of the all-time champions of explosive seed dispersal does just that. The aptly named dynamite tree (*Hura crepitans*) can be found growing naturally throughout much of tropical South America and is becoming invasive in parts of Africa as well. It is a tree you really don't want to mess with. Its sap is toxic, and its bark is covered in nasty looking spines. So, not only is this tree well-defended, it is also extremely effective at getting its seeds out into the environment. The dynamite tree produces baseball-sized fruits that look an awful lot like small, bland pumpkins. Each ridge of the pumpkin-like fruit is made up of a carpel that contains the seeds.

Each ridge or carpel of the fruit is connected to its neighbor by a thin layer of tissues. As the fruit matures, it gradually begins to dehydrate. As water is slowly lost from its tissues, they begin to contract. The more the fruit contracts, the greater the stress on the walls. Eventually, tiny fractures develop between each carpel.

These grow over time until, at last, the fruit can no longer hold itself together. With an audible snap, the whole fruit explodes, rocketing each seed-containing carpel outward from the tree at speeds of up to 156 m.p.h. (251 km.h.)! I would certainly not want to be standing near the tree without some sort of protective gear when this happens. As you can imagine, such speeds can carry seeds quite a distance. One experiment found seeds could travel as far as 147 feet (45 m.) from the parent tree, which is good news for seedlings that don't want to grow up in the shade of their mother. The dynamite tree has definitely earned its common name.

Examples of ballistic dispersal like these are surprisingly common among plants. Shooting your seeds out into the environment with force is a great way to ensure that they at least stand a chance at ending up in a favorable spot for germination. Still, there are no guarantees in nature, especially when the laws of physics are the decider. After all, plants are not expert marksmen. As effective as they can be, all forms of ballistic dispersal come down to chance. Moreover, environments can change, habitats become fragmented, wind isn't always available, etcetera, etcetera. Propagule dispersal is tricky yet essential, which is why numerous other species have evolved more directed means. Among the most interesting and effective are those that, like pollination, involve animals. Dispersal by animals is called zoochory, "zoo" obviously referring to the animal portion, and "chory" meaning plant propagule dispersal. Here, again, space will limit me to only a handful of my favorite examples, but know that there exists seemingly endless variation on the theme.

In Defense of Plants

Animal dispersal, despite all its variation, comes in two major forms. There are those plants that equip their propagules in some way so that they stick to the fur, feathers, or skin of an animal. We call this epizoochory. Since these plants do not require an animal to consume their propagules, they generally aren't fleshy or appetizing in any way. Instead, they usually come armed with a means of sticking to a passing target. They can be sticky or gooey or covered in barbs and spines. The key to epizoochory is that at some point, the propagules get dislodged, hopefully without being destroyed in the process.

The other form of animal dispersal involves the consumption of propagules, usually with the help of some form of tasty morsel attached to or surrounding them. We call this endozoochory because the propagules are moved around the environment as they pass through the gut of whoever ate them. There are added advantages to endozoochory beyond dispersal, too. For one, the propagules often benefit from passage through the gut as powerful stomach acids effectively break down any physical barriers containing the embryo, making germination much easier. Also, once they have passed through the gut, the propagules are often deposited in a bundle of fertilizer, aka poop.

Endozoochory should be one of the most familiar things in the world to our species. We take part in it every time we eat raspberries, blueberries, or any of the other myriad fruits that humans enjoy. Now, to be fair, most of us no longer function as effective seed dispersers. Our poop ends up in the sewer, far away from anywhere remotely suitable for germination. Today, most of the seed dispersal for commercially important fruits happens with the farmer. Still, in producing tasty, nutritious fruits that humans enjoy, some plants have greatly benefitted from our activities. Fruits, after all, evolved as a means of attracting the attention of potential seed dispersers. Most of the time, however, this does not involve humans. Attracted by bright colors and the promise of a meal, animals of all kinds gobble up fruits and the seeds they contain. In most cases, the edible parts of the fruit are digested and the seeds, armed with seed coats able to resist stomach acids, pass through the gut and exit out the back end. Usually, this occurs sometime after the fruit was initially consumed, meaning the seeds have traveled some distance before they are "released" back out into the environment.

Among the most famous seed dispersers are birds. Birds love fruits and seeds, gobbling them up with relish. Though some birds certainly destroy seeds, many more make it through the gut with no issue at all. One reason birds are so effective at dispersing seeds is because they can fly. Few other animals can so easily cover such distances as birds and the fruit eaters carry a load of seeds in their gut wherever they go. Most of you reading this right now will have no trouble finding examples of bird-dispersed plants around your neck of the woods. Take a walk outside and look around at all the wonderfully colored, bite-sized fruits available. In fact, with a trained eye and attention to detail, the role of birds in dispersing numerous plant species can be very obvious. Here in eastern North America, some of the best examples of what I am talking about can be seen in species like the black cherry (*Prunus serotina*).

Every summer when the fruits of the black cherry in our yard ripen up, dozens of birds descend upon it day after day until nearly nothing is left. A similar spectacle likely plays out on every black cherry throughout our town. Birds are so quick to gobble them up that we barely get a chance to sample some for ourselves. The birds get a nutritious meal, and the trees are spreading their seeds much farther than if they were to simply fall off the tree. There is a hiking spot near where I grew up that was a small farm only a century ago. Today, it is a nature preserve, and the forest has regained part of its former glory. Still, despite nearly a hundred years of regrowth, evidence of the old farm still abounds. One of my favorite clues to its prior existence is a straight row of mature cherry trees that runs through a section of the forest. They seem so unnatural, standing in single file in an otherwise stochastic assemblage of broadleaves and conifers. The spectacle is made all the stranger in that all the cherry trees appear to be roughly the same age and size. How did this happen?

The answer is a bit more natural than the scene implies. Birds are the foresters in this situation. That line of cherry trees demarcates the long-gone border of a pasture. A fence once stood where the cherry trees are today. I can picture birds flying out from the small patch of forest nearby and perching along the fence as they preened in the warmth of the late afternoon sun back when the property was a farm. Year after year, birds would repeat the process, and each one of them likely relieved themselves as they perched. Their droppings contained, among other things, the freshy cleaned and scarified seeds of black cherry trees. Each time a bird pooped, the seed within found itself in a favorable spot for germination and growth. While the farm was still in operation, the farmer likely did what they could to remove those trees. Not only would they damage the fence but, like most cherries, they pumped cyanide into their leaves and stems as a form of defense against herbivory. Cows and horses, having not evolved with black cherry trees, do not realize this and can be fatally poisoned if they each too much of them. One day, the farm stopped operating. The land changed hands and became the nature preserve it is today, and the forest began to take back the land that rightly belongs to it. All those black cherry seeds deposited along the fence were no longer being dug up or cut back, and they sprang up with vigor. The end result was a straight line of cherry trees in the middle of an otherwise natural-looking forest.

Fruit-eating bats are also incredible seed dispersers. This is especially true in the deserts of the Americas where for many cacti, the same species of bats that pollinate their flowers also disperse their seeds. One of the best examples of this can be found in the Tehuacán Valley of Mexico. This region is known for its high diversity of columnar cacti. Entire forests are comprised of their spikey, succulent stems. Bats are so important for dispersing cactus seeds in these forests that experts believe that the fruits of species like *Stenocereus weberii*, *Stenocereus pruinosus*, and *Mitrocereus fulviceps*, have taken to mimicking the color and odor of cactus flowers as a means of attracting the attention of bats like the lesser long-nosed bat (*Leptonycteris curasoae*). The efficiency of bats at dispersing seeds of the cacti that comprise these forests makes them vital for the existence of these habitats. However, the contribution of bats isn't restricted to arid regions of the world.

In the tropics, bats are cited as one of the main reasons why some trees are able to recolonize forest clearings after disturbance or logging. Trees that excel in disturbed areas such as those belonging to genera like *Cecropia*, *Solanum*, and *Vismia* produce large fruits full of seed-laden pulp that fruit-eating bats can't resist. Bats have surprisingly fast digestive systems, and some species will poop out a meal only twenty minutes after consuming it. Often, this is done while in flight. As bats cross patches of open land on their way to new patches of forest, they poop out all the seeds contained in their last meal. This seed rain is what will form the foundation of the forest that will spring up to replace what was recently lost. Unfortunately, our need to light up the night is keeping bats out of areas that desperately need an influx of seeds to regenerate. Bats are significantly less likely to spend time in areas with heavy light pollution, especially the type of lighting used on streets, buildings, and houses. This means fewer seeds end up in these areas. Essentially, as humans expand into these forested habitats and begin lighting up areas at night, fewer bats hang around and the forests suffer the brunt of this absence.

Whether dispersal occurs on the outside or the inside of an animal, zoochory is an extremely effective means of spreading potential offspring out into the environment. The process always reminds me of a quote by Terence McKenna, "animals are something invented by plants to move seeds around." Though I like the sentiment Terence was going for, this isn't entirely accurate. Animals evolved in the ocean long before plants evolved from algae and even longer before seeds evolved. And yet, he wasn't too far off the mark with the idea that plants shape animals just as much, if not more, than animals shape plants. Like we discussed with pollination, natural selection shapes both animals and fruits in tandem, rewarding successful genetic "experiments" with reproduction and "penalizing" those that are unsuccessful with death.

Whereas we know quite a bit about the role animals like birds and bats play in this process, there is another group of seed dispersers that are proving to be vital to the long-term health and survival of tropical forests around the globe: fish. The idea of seed-dispersing fish may come as a shock to some, but mounting evidence is showing that fruit-eating fish play a major role in the reproductive cycle of many tropical plants. This is especially true in seasonally flooded tropical forests. To date, more than a hundred different fish species have been found with viable seeds in their guts. In fact, some fish species like the pacu (*Piaractus mesopotamicus*) specialize in eating fruits.

Take, for instance, the tucum palm (*Bactris glaucescens*). Native to Brazil's Pantanal, this palm produces large red fruits, and everything from peccaries to iguanas readily consume them. However, when eaten by these animals, the seeds either don't make it through the gut in one piece or they end up being pooped out into areas unsuitable for germination. Only when the seeds have been consumed by the pacu do they end up in the right place and in the right condition. It appears that pacus are the main seed dispersal agent for this palm. The tucum palm isn't alone, either. The seeds of myriad other plant species known to inhabit these seasonally flooded habitats seem to germinate and grow most effectively only after having been dispersed by fish. Even outside of the tropics, scientists are discovering that fish like the channel catfish (*Ictalurus punctatus*) are important seed dispersers of riparian plants such as the eastern swamp privet (*Forestiera acuminata*). Without fish, these plants would have a hard time with seed dispersal in their seasonally flooded habitats. Lacking a dispersal agent, seeds would be stuck at the bottom of a river, buried in anoxic mud. As fish migrate into flooded forests, they can move seeds remarkable distances from their parent plants. When the flood waters recede, the seeds find themselves primed and ready to usher in the next generation.

On the other side of the zoochory coin is epizoochory, dispersal that doesn't involve seeds being eaten. Plants have evolved numerous means of doing this including mucilage, spines, burs, and sticky hairs. And just like endozoochory, it's a very effective means of dispersal. Many of us can empathize with the thought of having to remove burs from a dog's fur, our own hair, or articles of clothing. My childhood dogs used to come running into the house completely covered in the burs of burdock (genus *Arctium*) and bedstraw (genus *Gallium*) after an afternoon of roaming my parents' property. Examples such as these are the most obvious, but the world of epizoochory is much broader than our backyards and involves some surprising strategies and players.

The hooks on burdock burrs are formed from highly modified leaves called phyllaries.

If you have ever tried growing a Chia Pet or just soaked chia seeds in water, you will know just how sticky some seeds can get. That is why they are able to stick to the terracotta figurines that form the base of your beloved Chia Pet. The seeds of chia (*Salvia hispanica*) are but one example of seeds that turn gooey when wet and, believe it or not, that mucilage aids in their dispersal. Mucilage can get stuck on everything from fur to feathers, and even scales. All such seeds need is water. Chia seeds aren't nearly as sticky when dry, which means that they won't get moved around during long periods of drought. Only when the landscape is damp with rain does this seed dispersal strategy come into play. Rainy periods also signal an uptick in the movements of many desert species, which increases the likelihood of seeds coming into contact with a potential vector. Once the seeds fall off whoever is carrying them, that goo also helps them stick to whatever substrate they land on, increasing their chances of germination.

One of the most important and overlooked forms of epizoochory involves ants. Ants are frequent friends of plants and moving their seeds around is but one of the many ecological services ants provide. Dispersal via ants is known as myrmecochory and it involves some anatomical specialization on the part of the plants involved. To attract the ants' attention, some plants attach fleshy appendages to their seeds called elaiosomes. Elaiosomes come in a variety of shapes, sizes, and even colors but they are generally packed full of lipids and proteins. Some of them even emit unique odors that are thought to entice ants. Foraging ants take these seeds back to their colonies where the elaiosome is eaten and the seed is discarded. Ants have special chambers in their colonies for trash. They are basically little underground compost heaps.

Discarded in this way, the seeds find themselves in a very stable, nutrient-rich area where they can safely germinate. The ants get a little meal, and the plant has provided its offspring with one of the safest germination spots in the surrounding environment. There is also some evidence to suggest that the seeds gain a cleaning benefit from the ants. Like the Karner blue caterpillars, it is possible that ants, with their antimicrobial fluids, may inadvertently clean seeds that enter their nest. Because fungi and other microbes are one of the leading causes of seedling mortality, it is very possible that this is yet another added benefit of having ants as your seed dispersal agent. The sheer number of plants species that utilize ants in this way is staggering. Here in North America, the majority of myrmecochorous plants are spring flowering plants like violets (genus *Viola*), trilliums (genus *Trillium*), the aforementioned wild ginger (*Asarum canadense*), trout lilies (genus *Erythronium*), and bloodroot (*Sanguinaria canadensis*), just to name a few. There is generally a lot less food available to ants in the spring, making the seeds of these species very appealing. Once summer hits, scavenging ants are less likely to pay attention to seeds in lieu of more nutritious foods.

Ant dispersal is not limited to the temperate zone either. It occurs in the tropics as well and it is likely that many more examples will be discovered in the coming years. One of the most remarkable examples in my mind involves a group of orchids known as bucket orchids (genus *Coryanthes*). These orchids grow from the branches of trees but not without the help of ants. You will only ever find a bucket orchid growing in what is referred to as an "ant garden." Throughout their range, bucket orchid roots provide the scaffolding on which specific groups of tree-dwelling ants build their arboreal nests. Ants actively search out the surprisingly fleshy seeds of bucket orchids and bring them back and plant them in their nests so that they will always have new scaffolding on which to expand the colony. Moreover, because ants viciously guard their homes from intruders, the ants act as bodyguards for the orchids, tending them and cleaning them as they grow. Few wild plants receive better care from other organisms as bucket orchids do from their resident ant colony. The lives of these two completely unrelated organisms are intricately tied together, and it all starts with seed dispersal.

Plants with seeds aren't the only ones to utilize animals to disperse their propagules. That is the main reason why I have been so particular about using the term propagule throughout this chapter. Some mosses have entered into specific relationships with animals as a means of dispersing their spores as well. One of the coolest examples of this involves a genus with the less-than-endearing name of the poop mosses (genus *Splachnum*). As the name suggests, these mosses are specialists on animal poop. What's more, since different animals produce different kinds of poop with different textures and chemistry, different species of poop moss specialize on different kinds of poop. Deer poop is a favorite haunt of *Splachnum ampullulceum* whereas coyote or wolf poop will be inhabited by *Splachnum luteum*.

The problem with poop is that it decomposes quickly and that means that poop mosses can't leave spore dispersal up to chance. Instead of relying on wind, these mosses utilize another group of poop specialists: flies. Poop moss spores are produced at the tip of very long, very ornate sporophytes. They are large, colorful, and some even produce a fetid odor in hopes of attracting their winged spore dispersers. When a fly lands on the sporophyte to investigate, it gets covered in sticky poop moss spores. When it eventually lands on another pile of poop, some of the spores will fall off and begin a new moss colony.

Animal dispersal doesn't even need to involve seeds or spores to be effective at helping plants conquer new territory. It can even involve pieces of the plant itself. Some plants are capable of growing entire individuals from only a piece of living tissue. Sometimes this is a branch or a leaf, other times it's an entire stem. I am especially fond of the jumping cholla cacti (*Cylindropuntia fulgida*) for this reason. The name jumping cholla strikes fear in the heart of many a Westerner for their supposed ability to jump out and attack anyone unlucky enough to get too close. Each of the cacti's cylindrical pads is brimming with extremely sharp spines. Even the slightest amount of pressure will lodge them into your skin.

In reality, jumping cholla do not jump. The common name comes from the fact that the individual stem segments of the cactus are only weakly attached to one another. The slightest touch will detach them, giving unfortunate hikers and horseback riders the idea that the plant has leapt onto them. Each cholla spine is covered in backward pointing barbs, which ensures that they stick fast into whatever they have managed to puncture. Those backward pointing barbs also makes removing them a painful experience. If the unfortunate passer by happens to lack opposable thumbs to help in removal, there is a good chance that those cactus pads are going for a ride, and that is exactly what they are looking for.

Looking at those spines, it is easy to see why jumping cholla is
so good at hitching a ride on unsuspecting animals.

Jumping cholla utilize this strategy as a form of vegetative reproduction. Stem segments that stick to passing animals can be carried far from the original plant before they are successfully dislodged. If the cholla is lucky, this will happen in spot that is favorable for growth and thus a new plant can begin life in new territory. The result of this is a rather quick lifestyle for this cactus. Stands of cholla can be relatively short-lived. There is a lot of competition for light and space in mature cholla stands. Being able to get as far away from that congestion is very beneficial in the long run. This mobile form of vegetative reproduction has worked so well for the jumping cholla that an entire mythos has developed around their dreaded "jumping" habits.

The various seed dispersal mechanisms described above are but a mere sample of the diversity of strategies plants have evolved to overcome the fact that they are rooted in place. However, whether it's wind, explosions, shots, guts, fur, feathers, or skin, seed dispersal for the vast majority of plants is largely a game of chance. Not all propagules are going to be successful. In fact, nearly all spores and seeds released into the environment are doomed to fail. That is one reason why plants make so many over their lifetime. Successful reproduction for most plant species is a numbers game. However, there are a few plants out there that have taken seed dispersal to a completely new level by physically planting their own seeds. In doing so, these plants have eliminated as much chance from the seed dispersal process as possible.

It may be a surprise learn that some plants do their own gardening. The idea of a plant digging into the soil and depositing a seed seems outlandish and indeed, it is a very rare form of seed dispersal. To date, only a handful of examples have been described. Interestingly, one of the most famous plants in the world just so happens to fall into this category. Millions of tons of its seed are produced each year to be consumed in sandwiches, on airplanes, at ball games, and so much more, and yet we give almost no thought to the plant that produces them. I am, of course, talking about the humble peanut (*Arachis hypogaea*).

The act of a plant producing or even placing fruits underground is referred to as geocarpy. There are a few different ways in which geocarpy happens. For peanuts, the act of planting its seeds occurs following pollination. Upon fertilization, the flower stalks elongate and the stems carrying them bend toward the ground. Once they touch soil, the flowering stalks push the developing seed pod underground. Only then will the pods swell and mature into what we recognize as an unshelled peanut. In the wild, the seeds would germinate underground, never at the mercy of the vagaries of other forms of seed dispersal. In fact, peanut seeds are incapable of germinating aboveground as the embryos they contain only become active in a dark, subterranean environment.

Geocarpy is especially useful for plants growing in difficult habitats like cliffs or rock walls. Seed dispersal in such habitats is a challenge as seeds taken away by wind or animals are unlikely to end up in a crack or crevice suitable for germination. It makes sense then that a more guided approach to seed dispersal would help ensure that seeds have a chance. Such is the case for the ivy-leaved toadflax (*Cymbalaria muralis*). Originally native to Mediterranean Europe, this unassuming yet pretty plant now enjoys a near global distribution. Humans are part of the reason for this as ivy-leaved toadflax has reached minor celebrity status in gardening circles over the years. Though humans certainly got the ball rolling, much of its spread is due in part to its propensity for planting its own seeds. The means by which it does this are fascinating and well worth a closer examination.

The flower stalks of the ivy-leaved toadflax exhibit phototropic behavior, meaning that they respond to the direction of incoming light. During its flowering stage, the flowering stalks initially exhibit positive phototropism, growing toward the direction of the light. This ensures that the flowers will be exposed to the sun's rays and readily available to pollinators. Once pollination occurs, a change takes place within the flowering stalk. As the ovary swells with developing seeds, the stalk becomes negatively phototropic. This causes it to grow away from the direction of the light. Because ivy-leaved toadflax is a cliff dweller by nature, negative phototropism means the flowering stalks tend to grow toward and eventually into dark, damp crevices. This is exactly the kind of spot that ivy-leaved toadflax seeds need to germinate. Not only are crevices sheltered and largely free of seed predators, they are also among the only surfaces on cliffs and rock walls where tiny amounts of soil and moisture can collect. By avoiding the light and sticking their fruits into dark holes, the ivy-leaved toadflax gives its offspring a fighting chance at survival in their vertical habitat.

Ivy-leaved toadflax is most at home growing out of cracks in cliffs and rock walls.

Now, before we end this chapter, there is one species I would be remiss not to mention. The plant I have in mind lives in the rainforests of Central and South America and goes by the common name of walking palm (*Socratea exorrhiza*). If there was ever a plant that could be said to "walk" around the landscape, this would surely be it. Before I get ahead of myself here, I think it is worth dispelling any image of an actual walking tree. These palms are by no means Ents. However, their biology does provide them with more freedom than their more firmly rooted cousins and I feel that they deserve honorable mention.

One does not soon forget their first encounter with this odd palm. For me, it took place on a backpacking trip through Central America. On a hike through a chunk of rainforest in southern Costa Rica, we came to a hillside with surprisingly little vegetation, which gave us an uninterrupted view of the trees. Growing on that hill were two of the strangest palms I had ever laid eyes on. Instead of a single stout trunk, these palms were balanced on what looked like a dozen or so spikey tentacles. Our friend and guide for the week, plant guru Dave Janas, told us that we were looking at a pair of walking palms and those "tentacles" were their roots. Whereas most palms invest in heavy trunks, the walking palm sends out a lanky set of stilt-roots upon which the palm grows like some otherworldly arborescent squid.

Decomposition is as common as it is rapid in tropical forests. Due to all the heat and humidity, the omnipresent threat of rot means there is a constant rain of limbs and snags from the canopy above. Trees regularly topple as well. For most plants, getting flattened by such debris is usually fatal. However, for the walking palm, getting toppled by a fallen limb isn't necessarily bad. It has been observed that walking palms flattened by fallen debris can actually "walk" themselves out from underneath. The key to this lies in those stilted roots. Most of the trunk of a walking palm can produce stilt-roots and when one gets knocked over, the trunk can swing into gear, sending out new roots into the soil. If this happens quick enough, the walking palm can find itself rooted into a new spot free of the weight of the branch that squashed it. From there, the walking palm can continue its journey into the canopy, leaving the remains of its flattened trunk behind.

Getting squashed isn't the only danger walking palms face in a dense rainforest. Light is at a premium in the deep shade of a rainforest understory. Plants grow so quickly in the tropics that a hole in the canopy that provided ample light one week may quickly close in the next. If a plant were able to "move around" it could potentially relocate to a sunnier spot. In a sense, this is what the walking palm does. Like the flowering stalk of the ivy-leaved toadflax we just met, walking palm stems are positively phototropic, always leaning toward a light source. Leaning can put so much stress on a trunk that some trees eventually topple over. The walking palm gets around this by sending down those stilt-roots, which provide support as it chases light through the understory. Once in a favorable location, the walking palm will right itself and continue upwards. In a sense, this palm "walks" itself around the forest in search of the best light. Once in position, the old trunk and roots will often rot away. For a young tree, this process can happen in as fast as two or three years. Whereas most trees are stuck where they germinate, the walking palm has, in a sense, freed itself from such restrictions.

Sure, plenty of arguments could be made that this form of "walking" is no different than what is observed in any plant capable of vegetative growth. Whether via rhizomes (underground stems) or stolons (horizontal stems that lay on the surface of the soil), many plants can forage for better territory. I agree to a point, however, I think the walking palm has taken this type of growth to the extreme. It also provides the most striking visual example of plant foraging abilities.

My goal for this chapter was to demonstrate that plants are not the static organisms we make them out to be. They are fully capable of moving around the landscape; they just do it in ways that we don't readily recognize or perceive. It is also worth the reminder that plants are functioning in a habitat full of other life forms. No plant, and indeed, no organism, for that matter, operates in a vacuum. Plants must constantly compete with one another for space. They also must defend themselves from being eaten. Plants have evolved countless mechanisms for competition and defense. Some of their abilities are so effective as to seem almost sinister. I can't reiterate enough that such petty human emotions are absent from the living world, but that doesn't mean that plants don't do what is needed to survive long enough to reproduce. In the next chapter, we will take a closer look at how plants fight for survival in a world devoid of care and concern.

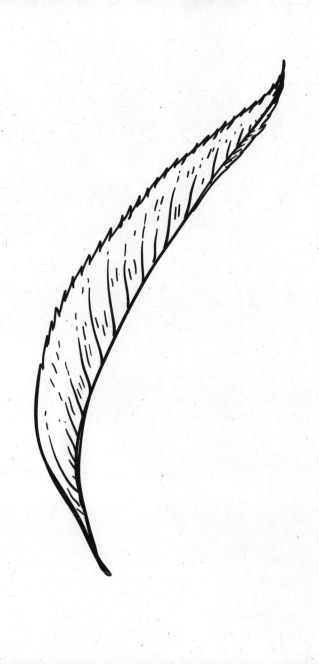

CHAPTER 5

The Fight for Survival

must admit that there is a considerable part of me that enjoys shattering misconceptions. Mind you, I don't fancy myself a troll. I would never bully someone for their beliefs, but in a world full of misinformation, I feel it is important to dispel as many myths as possible. One of the biggest myths that people must overcome is that idea that nature is full of peace, harmony, and balance. These concepts are human inventions and are applied to the world so that our feebly conscious minds can deal with the unfeeling, uncaring essence of nature and the universe as a whole. Just as the phenomenon known as pareidolia causes us to see faces in the rock formations on Mars or the Virgin Mary on a burned piece of toast, our need to apply meaning to the world causes us to see altruism in what are essentially cases of selfish give and take, and harmony where it doesn't exist.

Distilling ecological science is a balancing act. One must strive to make organisms like plants relatable to an audience without imbuing them with undue intent or anything approaching consciousness as we know it. Anthropomorphizing is excusable, to an extent. I am doing what I can to walk the line between helping people understand how plants make a living without making them out to be too much like us. Plants are nothing like us, which is part of the reason why I like them so much. Nonetheless, I find that drawing similarities to war in this chapter is appropriate. We define war as armed conflict between different nations, states, or groups, and I feel like it is a fitting metaphor for the myriad ways plants compete with one another and defend themselves. At the heart of any human war is control of resources. For humans, this usually means land, wealth, or oil. For plants, resources can mean a lot of different things; water, nutrients, space, light are all chronically in short supply and plants are doing what they can to maintain their positions on the landscape and obtain what they need to grow and reproduce.

As you are going to learn, some of the ways in which plants go about competition and defense would bring about severe sanctions from the United Nations if we were to try them. It is important to always remember that plants are living organisms that are struggling to survive. Due to their sessile nature, natural selection has resulted in the evolution of some incredible botanical defense strategies. We will start by looking at competition because the idea that plants interact and compete with one another was such an important realization for me in shaping my plant obsession.

All species experience competition to some degree. In the botanical world, competition usually boils down to space and light. Plants need room to spread their roots and absorb nutrients and water. Soil is a complex and dynamic environment, but there is only so much room available. The denser the roots of one plant, the less room there is for the roots of others. This is why good planting instructions come complete with spacing recommendations. Plants also need space aboveground to spread their leaves. When the leaves and branches of one plant overtop another, the larger of the two gets more sunlight while the smaller slowly starves. It is entertaining to picture a quiet forest in this context. Like people on a crowded bus, plants are always jostling for the space they need; they just do it without making noise.

One of the easiest ways in which plants overcome competition from their neighbors is by growing taller. We are used to seeing plants of various sizes, but have you ever stopped to think why there is so much variation in height among species? Why are there miniscule plants like whitlow grass (*Draba verna*) whose leaves top out at only a millimeter or two above the soil and towering goliaths like redwoods (*Sequoia sempervirens*) whose canopies reach over 370 feet into the sky? The answer to this is competition. From an evolutionary standpoint, no other explanation makes sense. Growing taller is not only costly, it makes plants more vulnerable. Investment in the kinds of sturdy tissues needed to grow a plant of any height is certainly higher than if a plant were to live out its life hugging the ground. Also, tall plants frequently fall victim to strong winds or lightning strikes. Growing tall also makes plants more vulnerable to drought. The taller a plant gets, the harder it is for water to move up through their tissues against the pull of gravity.

When you walk into a field full of lanky herbs and grasses or a forest full of towering trees, you are essentially witnessing an evolutionary arms race many millions of years in the making. It is a race that plants have been running since the Devonian, some 385 million years ago. As plants left the water and began covering the land, some inevitably grew taller than others. Getting above their neighbors meant more sunlight but it also meant increased access to pollen vectors like wind or insects and, by the same logic, greater chances for spore and seed dispersal, too. Once the race was started, there was no turning back. Now, I realize that not all plants are the size of redwoods. Plenty of species have done just fine living out their lives closer to the ground. However, this, too, has everything to do with competition. If a lineage of plants evolved a means of photosynthesizing with less light, they didn't need to grow as tall. Many species do just fine in shady environments. Similarly, if plants like the aforementioned whitlow grass do well in unforgiving habitats where even their neighbors stay small, then there is little evolutionary impetus to grow tall. If a plant can obtain the resources it needs while staying small, it doesn't need to waste energy trying to outgrow its neighbors.

Let's return to trees for a bit here as I feel like they provide us some of the best evidence for competition among plants. Trees have taken plant height to its extreme. Tall trunks, thick branches, a complex canopy—all that wood must be grown by the tree and most of it offers no additional benefit other than providing a taller pole on which to grow leaves. At face value, trees seem like the most lavishly wasteful forms of plant growth. That is, until you consider competition. If there was no competition, trees likely wouldn't exist at all. And yet, when we talk about trees and forests, our minds instantly conjure up images of peace and tranquility. Rarely, if ever, do people consider all the struggle and competition that led to trees in the first place. I think this is due in large part to recent scientific discoveries about forest ecosystems and the way in which the media has misinterpreted them.

If, at this point, you find yourself screaming into the pages of this book saying something to the effect of "But Matt, what about the wood wide web? Science says trees share with one another! They don't compete, they nurture," this is exactly what I am talking about. I'm not pointing fingers at anyone in particular, but there seems to be a tendency among journalists to draw unnecessary connections between human culture and nature in an attempt to make a thinly veiled point about the way we humans get along. A few years ago, a story caught fire in the media that introduced the idea of the wood wide web to the public. The idea was pretty incredible; trees partner with mycorrhizal fungi in the soil and can send nutrients and even communicate chemically through the vast network of fungal threads that connect their roots. Over the span of a few weeks, numerous outlets grabbed onto the story and ran with it with little regard for scientific accuracy. "Trees talk to each other! Trees are sharing with each other! Trees are taking care of their own!" All manner of claims were being made about altruism in trees.

So many plant species would not exist today if it were not for the relationships they form with these soil-dwelling fungi. Trees are no exception to this. In fact, given that mycorrhizal relationships are found throughout all plant lineages, from nonvascular to flowering plants, one could argue that this may be one of the most important mutualisms on our planet. There are a lot of variations on the theme of mycorrhizal relationships, but all of them come down to simple give and take. The fungi provide the plants with access to nutrients that would otherwise be unavailable, while the plants give back in the form of carbohydrates made during photosynthesis. Thanks to advancements in technology, we are gaining insights into just how complex these relationships can be. Recent work has shown that indeed, nutrients can be shared among trees thanks to the network of fungi connecting their roots. Evidence also suggests that some form of communication can in fact happen through these fungal networks, with hormones and pheromones signaling threats like that aphids or some other pests are present. It would seem that rather than competing with one another,

In Defense of Plants

trees are using mycorrhizal fungi like the internet, warning each other of pending disaster and transferring resources to those in need. It's as if forests are one giant commune with all the residents living in harmony. I'm not convinced of this view.

Let me be clear, I find nothing wrong with the scientific side of this story. I have read the research and I think it stands on solid scientific ground. What I am not convinced of are the subsequent interpretations. In our never-ending quest to give meaning to everything in the universe, humans never seem to miss an opportunity to humanize life in all its various forms. Yes, there is evidence for nutrient transfer and communication between trees via the fungal network, but everyone seems to forget the role of fungi in all of this. Sharing among trees across the mycorrhizal network makes a lot more sense when you think about it from a fungal perspective. Mycorrhizal fungi are living organisms, too, and they rely on trees for their carbon needs. Without a thriving network of trees to provide them with carbohydrates, they are left to their own devices. If the fungi need the trees for food, why would they let members of their network fail? From my perspective, fungi are the ones overseeing all the sharing. They stand to gain a lot from a thriving network of healthy trees.

Even if we remove fungi from the equation, evidence suggests that sharing among trees most often occurs between related individuals. Across all walks of life on our planet, we tend to see more "altruistic" behavior when organisms share DNA. Parents care for their own children more than the children of others; siblings care more for each other than non-siblings. We refer to this as kin selection. Simply put, life is DNA and DNA "takes care" of its own copies. Sharing among related trees makes sense because offspring are close genetic relatives of their parents. Also, circling back to tree height and vulnerability, one must also realize that forests create their own microclimates. Their density and shade mean forest understories are generally cooler, wetter, and less windy. Every tree that dies creates a gap in the canopy, which in turn alters the local microclimate. When trees die, those living around them become more vulnerable to drought and wind. The game of life and death forces some level of cooperation among organisms within a community whenever conditions are tough and trees are no less a part of that game than any fungus, beetle, or bird. To ensure the community in which you live survives is to ensure your own future as well. But alas, I digress.

Growing tall is one way to outcompete your neighbors, but it's not the only option. Remember the garlic mustard example in Chapter 2? Garlic mustard is not the only allelopathic plant to have evolved chemical means to stunt the germination, growth, and reproduction of its competitors. Myriad other plants use this strategy as well. Take, for instance, the case of the sandhill rosemary (*Ceratiola ericoides*), which lives in sandy habitats in the extreme southeast of the United States. This member of the heath family uses allelopathy not only to reduce competition, it also utilizes it as a means of avoiding fire. The sandhill rosemary lives in scrubby habitats that are prone to regular wildfires. However, unlike most of its neighbors, sandhill rosemary is not well adapted to cope with fire. These shrubs can be killed by even a modest burn and instead rely on a buildup of their seeds buried in the soil for regeneration. That doesn't mean that mature shrubs are completely vulnerable any time a fire sweeps through. The sandhill rosemary also utilizes chemical warfare to increase its chances of survival.

As anyone who has tried to light a fire knows, you need fuel. In the wild, plant materials like dry leaves, needles, and branches make up the fuel load. The more plant material lying around, the more fuel the fire can burn through. Much of the understory of sandhill rosemary habitats are filled with fire-adapted plant species. Grasses are among the most fire-tolerant of them all. Such grasses often release specialized compounds when they burn that cause fires to burn even hotter (yet another form of competition-reducing chemical warfare). This is bad news for species like sandhill rosemary. If grasses and other fire-adapted species grow too close, they not only increase the chances of fire reaching the shrub, but also increase the intensity of the flames. This is where the allelopathy comes in.

Allelopathic compounds released by sandhill rosemary inhibit the germination and growth of other plant species. This is especially true for fire-adapted grasses. Whenever sandhill rosemary drops its leaves, they decompose and release their chemical cocktails into the soil immediately surrounding the shrub. This keeps plants at bay in the immediate vicinity while their roots go a bit further. Since the roots of sandhill rosemary branch outwards from the plant, the effectiveness of its chemical warfare is increased by a radius of a few meters around each shrub. Stand back and look at a few of these shrubs and you will see that each is ringed by a few meters of bare sand. Indeed, it is believed that sandhill rosemary is keeping the surrounding area clear of most fire-adapted vegetation, thus increasing its chances of survival whenever fire sweeps through.

As effective as they can be, allelopathic chemicals take energy to produce. The more resources invested in chemical toxins, the less there will be available for growth and reproduction. This is probably why evolution has equipped some plants with a means of taking advantage of toxic substances already present in their environment. Such is the case for the common blackberry (*Rubus allegheniensis*). If you have visited the forests of eastern or western North America, there is a good chance you have encountered this species. Common blackberry can form dense, thorny stands wherever it grows, and excels at crowding out potential competitors. Instead of producing a toxic chemical cocktail to do this, the common blackberry uses toxic heavy metals already present deep in the soil.

The common blackberry has this amazing ability to translocate manganese from one layer of soil to another using its roots. This may seem like a funny talent for a plant to have, but it all becomes clear when you realize the effect it has on surrounding vegetation. Manganese can be very toxic to plants, especially at high concentrations. Common blackberry is immune to the damaging effects of this metal and has evolved two different ways of using manganese to its advantage. First, as mentioned, it redistributes manganese from deeper soil layers to shallow soil layers using its roots as a conduit. Second, it absorbs manganese as it grows, concentrating the metal in its leaves. When the leaves drop and decay, their concentrated manganese deposits further poison the soil around the plant. For plants that are not immune to the toxic effects of manganese, this is very bad news. Essentially, the common blackberry eliminates competition by poisoning its neighbors with heavy metals.

Sandhill rosemary shrubs poison the soil around them,
keeping competing vegetation at bay.

Often the biggest threats to plants come not from other plants, but from organisms that want to eat them. There are a variety of ways in which plants can fend off attacks from herbivores looking for a leafy meal. I like to place these methods into three distinct categories. There are physical defenses like thorns, prickles, and spines. There are chemical defenses that, like allelopathy, aim to dissuade or even poison herbivores. Finally, there are mutualists that act as bodyguards. Examples of all three of these methods could fill volumes, but even a cursory glance at a handful of examples paints a picture of plant defenses that will hopefully change the way in which you look at our botanical neighbors. As I have said countless times up to this point, plants are far from static, helpless backdrops to the rest of life on Earth. Plants are actively fighting for survival just like the rest of us.

If you had a childhood anything like mine, you probably fell into your fair share of bramble patches. I have "donated" so much blood to blackberries, raspberries, roses, and greenbriers that I almost feel like they could be running their own blood drives. Anyone who has had their flesh torn by the armaments on these plants can tell you how effective such anti-herbivore anatomy can be. Examples like these are so common that I don't want to take up precious space within this book reiterating that fact. Instead, I want to take a closer look at some physical defenses that are much less obvious to our eyes.

There are plants out there that combine physical defenses like these with some nasty chemicals. Probably the best example of this are the nettles. There are many different species of nettle, but where I live in eastern North America, you are most likely to encounter either stinging nettle (*Urtica dioica*) or the wood nettle (*Laportea canadensis*). Look closely at their stems and leaves and you will see they are covered in tiny hairs. These hairs are called trichomes, and they are what puts the sting in stinging nettle. Each trichome consists of an elongated cell that sits atop a pedestal. They are very brittle, and any contact will cause their tips to break. These hairs are also hollow. Once broken, each trichome functions like a mini hypodermic needle. The broken trichomes will penetrate the skin of any animal unlucky enough to brush up against them and inject an irritating fluid into the tissues of their "attacker." The fluid itself is quite interesting. Chemical analyses have revealed that it consists of a complex mixture of histamines, acetylcholine, serotonin, and even formic acid.

Not all herbivores are deterred by this defense. For instance, small invertebrates like caterpillars don't seem to have any issue navigating the stinging hairs. Indeed, nettles are important host plants for many different insect species. Instead, it is thought that the stinging nature of nettles evolved in response to large mammalian herbivores. This makes some sense as larger herbivores pose more of an immediate threat to the entire plant than do smaller invertebrates. Even more interesting is the response of some nettles to varying levels of herbivory. It has been found that heavily damaged plants will regrow leaves and stems with higher densities of stinging hairs than those of plants that have experienced lower rates of herbivory. This, too, makes a lot of sense. Stinging hairs and the chemicals they contain require resources to produce so plants that have not experienced high rates of herbivory do not bother allocating precious resources to their production. Also, in species like stinging nettle (*U. dioica*), female plants produce more stinging hairs than males. It is thought that since females must invest more resources into producing seeds than males do into producing pollen, they must also invest in more protection for these valuable reproductive assets.

Among the gnarliest combination of physical and chemical defenses involves tiny crystalline structures called raphides. Raphides are so effective at deterring herbivores that they have evolved independently in over two hundred different plant families. To find examples of plants that use this defensive strategy, most of you probably won't even need to leave your home. Raphides are present in varying amounts in common fruits and vegetables such as pineapples, kiwis, spinach, and chard. Thankfully, domestication has reduced their potent effects. Still, one should eat such foods in moderation as overdoing it can cause kidney problems. To find more potent examples, look over at your plant shelves. Raphides are particularly common form of defense in the aroid family (Araceae), which contains some of our most beloved houseplant species.

Take a bite out of a dumbcane (genus *Dieffenbachia*), pothos (genus *Philodendron*), or monstera (*Monstera deliciosa*) and it won't be long before your mouth and throat start to burn. Eat enough of it and your symptoms may also include intense numbing, oral irritation, excessive drooling, localized swelling, and possibly even kidney and liver failure (please, don't actually do this). Such symptoms are the result of a complex form of anti-herbivore defense strategy that starts with raphide crystals. Raphide crystals come in a variety of shapes and sizes but the nastiest of them all are shaped like tiny needles. Plants use calcium oxalate to build these crystals and they do so in specialized cells called idioblasts.

A *Deiffenbachia* or dumbcane growing in the shaded understory of a Costa Rican rainforest. Herbivores would be wise to steer clear of its toxic foliage.

Raphides are only the first part of the defensive equation. Within each idioblast, raphides are surrounded by acrid and toxic proteins. When plant tissues containing raphides are damaged, usually by chewing, the raphides shoot out of the idioblasts and into the oral cavity of the herbivore. This is where their needle shape comes in. All those microscopic needles cut into and tear the lining of the mouth, esophagus, and gut. The lacerations caused by the raphide crystals allows all those acrid and toxic compounds to enter the wounds. This is where things can get especially nasty. If the proteins are toxic enough, the herbivore now has far more to worry about than just the burning sensation. Smaller animals are more susceptible than larger animals, but severe organ damage and even death are not unheard of. As terrifying as this may sound, it should not scare you away from these plants. Provided you or your loved ones don't go nibbling on the leaves or stems, all will be fine. If anything, this remarkable form of plant defense should earn these plants even more respect than they already get.

Chemical warfare may be the most effective means of defense for plants. In fact, the botanical world is responsible for producing some of the most toxic compounds on Earth. It is incredible to think that such seemingly innocuous organisms like plants can be so deadly. Among my favorite plants to grow each summer are the jimsonweeds in the genus *Datura*. The beauty of these vespertine members of the tomato family obscures their deadly nature. The genus *Datura* (as well as many other members of the tomato family) are famous for their production of toxic alkaloid compounds like scopolamine and atropine. It would only take a small amount of these chemicals to completely ruin your week and slightly more to put you in a grave. Take a bite out of a *Datura* leaf and you may experience symptoms like convulsions, irregular heartbeat, dizziness, nausea, blurred vision, vertigo, and hallucinations so severe that you completely lose touch with reality. And that's only if you survive. I would say that more than any other plants, *Datura* remind me to use caution while gardening.

When it comes to defense, sometimes poisoning or killing herbivores outright isn't enough. Herbivores like caterpillars can be so numerous that eliminating one is akin to clipping your fingernails. They must be dealt with in their entirety or else the plant risks losing everything. That is one reason why scientists think a plethora of plants produce a compound called methyl jasmonate. Methyl jasmonate serves a variety of functions for plants. It is involved in important processes like germination, root growth, and fruit ripening, but it is also used in defense. Methyl jasmonate is often released when a plant is damaged. Neighboring plants can pick up on this compound and will begin to beef up their own defenses in response. Again, this is not altruism on the part of the victim. Rather, neighboring plants are likely eavesdropping on its distress to their own advantage. After all, if your neighbor is being attacked, there is a decent chance you will be, too.

One of the most interesting attributes of methyl jasmonate is that it can turn caterpillars into cannibals. Yeah, you read that correctly. Scientists studying methyl jasmonate in tomatoes discovered that this compound induced beetworms (a common agricultural pest) to turn on one another through cannibalism. Cannibalism in insects is not unheard of, even among the herbivorous species like beetworms. However, such cases usually only occur when all available food has been consumed and starvation sets in. It would appear that methyl jasmonate makes caterpillars resort to cannibalism much earlier than they would naturally. So, instead of eating the plant, the caterpillars turn on one another. Good thing it doesn't have the same effect on humans.

As effective as chemical defenses can be, they are nonetheless costly to produce. Compounds like alkaloids require ample resources like nitrogen that are often in short supply. Some plants get around this by waiting until a threat is near to start producing their defenses. Trees like the sycamore maple (*Acer pseudoplantanus*) and European beech (*Fagus sylvatica*) will only begin producing defense compounds like tannins after they have received a signal that their growth tips have been nibbled by mammals like deer. Via a complex chemical pathway, these trees are able to detect the presence of deer saliva in their wounds. When this happens, the trees begin to ramp up the production of tannins in their tissues, which bind to proteins in the animal's gut and slows digestion. As tannins increase, deer spend much less time feeding on leaves. By waiting until the threat is detected to produce their chemical defenses, the trees are saving valuable energy that can be used elsewhere.

But just as we saw with the common blackberry, if a plant can use its environment to do the work for it, why not avoid producing chemical toxins altogether? Such is the case for a class of plants termed hyperaccumulators. Hyperaccumulators don't fall into a single genus, family, or even order. A variety of plant lineages have evolved this strategy. What unites hyperaccumulators is that they all concentrate levels of heavy metals in their tissues that would be fatal to most organisms. Of course, such strategies only work if you live in a place with high levels of metal in the soil. Generally speaking, hyperaccumulators hail from regions of the world rich in metalliferous soils like serpentine.

Metalliferous soils are difficult for most plants to live in because of their naturally high metal content. The plants that do grow in such soils are often very restricted in their distribution and either cannot grow anywhere else or get outcompeted by other plants in less toxic soils. Hyperaccumulators have been found to take up a variety of metals including nickel, zinc, and cadmium. Some do this to such a degree that it actually changes the color of their sap. One of the best examples of a hyperaccumulator comes from a tree endemic to the island of New Caledonia called *Pycnandra acuminata*. Its sap contains so much nickel that it is the same blue-green color as the Statue of Liberty. New Caledonia is a hot spot for metalliferous soils so finding such a tree there is not terribly surprising. What is surprising is just how much metal this tree accumulates. One study found that its blue-green sap contains upwards of twenty-five percent nickel.

Though we can't say with one-hundred percent certainty, defense is the leading hypothesis as to why trees like *P. acuminata* toxify their sap. Whereas the high concentrations of heavy metals in their tissues are not toxic to the plants themselves, they are certainly toxic to anything that may want to eat them. Insects and other herbivores may be able to detect heavy metals within the tissues and will actively avoid feeding on those plants. If no other options are available, then eating such plants will poison the herbivores. One study found that locusts feeding on tissues containing high levels of heavy metals exhibited significant reductions in growth and development. As tantalizing as this evidence is, we still have a lot to learn about hyperaccumulation in plants.

Not all plants have the luxury of producing costly toxins or acquiring heavy metals from their environment for use in defense. For so many plant species, defense against being eaten doesn't come in the form of physical attributes or chemical warfare, but rather via fascinating alliances with other organisms. We have already been introduced to this idea back in Chapter 1 with the caterpillars of the Karner blue butterfly and again in Chapter 4 with the bucket orchids. In both cases, ants were recruited as bodyguards. These two examples are not isolated cases. Ants have teamed up with numerous other plants. Like the bucket orchids, some plant lineages take their relationship with ants to the next level by offering them a place to live. When it comes to colony defense, ants don't mess around.

One of the most familiar examples of plants that grow ant homes or domatia, as they are referred to in scientific circles, can be found in the genus *Tillandsia*. These wonderful epiphytic members of the pineapple family go by the common name of air plant because they get all their water and nutrients from rain and dust particles. Around thirteen species of *Tillandsia* have entered into a housing agreement with ants and the morphological adaptations needed for this make them easy to recognize. If you look closely at the leaves of these thirteen species, you will notice that they roll up to form tubes that

In Defense of Plants

lead down into their bulbous base. The space between the leaves forms a hollow chamber that provides a perfect microclimate for ant colonies. In many habitats, these *Tillandsia* offer better housing than the surrounding environment. One would be surprised at how many ants can fit in there too. Colonies containing anywhere between one and three hundred ants are not unheard of. Disturb a *Tillandsia* with a resident ant colony and you will quickly learn just how beneficial this relationship can be for the plant.

Some plants take this relationship to a whole new anatomic level. Two genera belonging to the coffee family (Rubiaceae), *Hydnophytum* and *Myrmecodia*, go beyond simply providing ants with a single chamber. They grow entire ant farms within their swollen bases, complete with multiple chambers connected by winding tunnels. Looking a cross section of the base of these plants is like looking at an idealized image of those plexiglass ant farms we played with as children. Unlike those ant farms, however, the accommodations provided by the ant plants are quite luxurious. Again, we see a situation in which the ant colony will defend their plant from threats of all sizes, but the plants get a bit more out of their cohabitation.

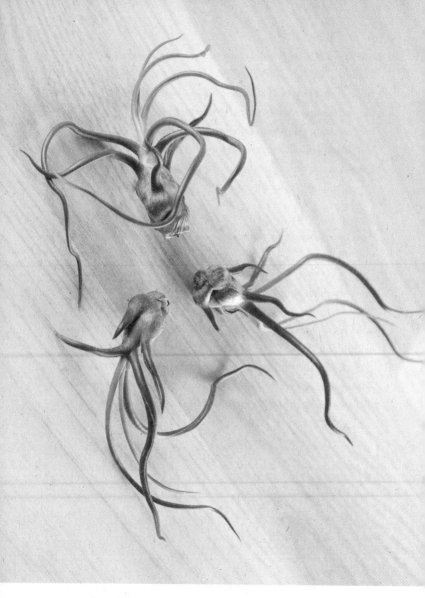

The swollen base of this ant plant (Myrmecodia sp.) is full of
tunnels and chambers where an ant colony can take up residence,
protecting it from herbivores and supplying it with nutrients.

Members of both genera grow epiphytically on trees, which means nutrients are hard to come by. Tree bark and debris just doesn't provide the same nutritional benefits as mineral soil. This is not an issue for ant plants as they have evolved a means of extracting the nutrients they need directly from their resident ant colonies. Ants are fastidious when it comes to keeping their colony clean and they will usually dedicate specific chambers within the plant for trash disposal. If you were to look closely at the walls of these chambers, you would notice that the ant plants furnish them with little nob-like growths. Those growths function like tiny roots and are capable of absorbing nutrients from the breakdown of all the waste that accumulates. So, not only do the ants provide protection, they also provide an excellent source of fertilizer for the ant plant.

Sometimes the relationship between ants and their botanical homes goes far beyond the individual plant. In the Amazon basin, there exists three genera of trees (*Duroia*, *Tococa*, or *Clidemia*) that provide ants with homes in exchange for complete dominance in the forest understory. Their resident ant colonies are so good at defense that they do not stop at fending off herbivores. These ants also attack and kill all competing vegetation. Locals refer to areas of the forest dominated by these trees as Devil's gardens. They are said to be the resting place of an evil spirit known to local tribes as Chullachaki. Anyone unlucky enough to stumble into his garden is said to be at risk of attack or curse. In reality, these gardens were created by ants. If ants encounter a seedling of their host tree while scouting, nothing happens. They go about their business and let the seedling grow into a future home. When they encounter a non-host tree, however, their behavior completely changes. The ants begin biting the stem of the plant, exposing its vascular tissue. As they bite, the ants also sting the foreign seedling, injecting minute amounts of formic acid into the wound. One or two ants isn't enough to bring down a seedling, but one thing ants have on their side are numbers. Not long after a foreign plant is discovered, an entire platoon of ants will descend upon the hapless seedling, stinging it repeatedly. In no time at all, the seedling succumbs to the formic acid injections and dies. By repeating this process any time a plant is found that will not eventually provide the ants with a home, the resident ants ensure no other plant will compete with those that do. Thus, a Devil's garden is formed.

Ants provide so many examples of defense mutualisms with plants, but they aren't the only players in the game. Some of the best plant bodyguards are so small that you would have trouble seeing them with the naked eye. Hearing the word "mite" as a gardener instantly makes me think of pests such as spider mites. This is not fair. The mite family is very diverse, containing countless beneficial species. Some mites are important predators at the micro scale, hunting down and consuming potentially harmful mite species or dining on disease-causing microbes and fungi. Their taste for pests has not been lost on some plants, and a surprising amount of plants such as red oaks (*Quercus rubra*), sugar maples (*Acer saccharum*), black cherries (*Prunus serotina*), and grapes (genus *Vitis*) go as far as to provide such mites with homes. Mite homes are produced on the leaves and often consist of little more than small divets lined with hairs. Despite their lack of furnishings, these divets function as important shelters for both the mites and their eggs. In housing beneficial mites, the plants are ensuring that they have a steady supply of hunters and cleaners living on their leaves. Predatory mites are voracious hunters, keeping valuable leaves free of microscopic herbivores while fungivorous mites clean the leaves of detrimental fungi that are known to cause infections such as powdery mildew.

Mutualistic relationships with mites must be very good for both plants and mites alike as they appear to have been cohabitating for a very long time. Recently, mite homes were identified in the fossilized remains of leaves dating back to the Late Cretaceous, some seventy-five million years ago. Even more exciting is that these fossils don't represent a single species of plant, but many. Plants and mites have been teaming up at least since *Tyrannosaurus rex* roamed the land. I am sure that future fossil discoveries will reveal these relationships to be much older than that.

I hope that the few examples I provide in this chapter are enough to convince you that plants are neither helpless nor peaceful. The struggle to survive is real in plants, and they are very good at it. It is always worth the reminder that plants made their way onto land roughly 470 million years ago, and they have maintained their dominance on the landscape ever since. That is extremely impressive for sessile organisms. It's kind of a shame that plants only get credit for their various survival mechanisms when we have discovered a use for them. Plants did not evolve in response to our needs, they evolved in response to the myriad pressures they face each day. Moreover, plants have not done this on their own. They have evolved an amazing variety of mutualistic relationships with other organisms, and each year, we uncover more examples.

However, the interactions between plants and the living world doesn't end at sex, competition, and defense. Some plants have gone to even greater lengths to get what they need to grow and reproduce. For some species, this means expanding their diet beyond what they can absorb from their roots. Across the globe, there exists plants that turn the tables on the whole "eat or be eaten" game by capturing and consuming animals. I am, of course, talking about carnivorous plants. Most of us will be familiar with the Venus fly trap (*Dionaea muscipula*) and some of us have probably grown a couple in our day, but the few carnivorous plants that have made their way into popular culture are just the tiniest tip of an incredibly strange, botanical iceberg. In the next chapter, we will take a closer look at just how amazing the world of carnivorous plants can be.

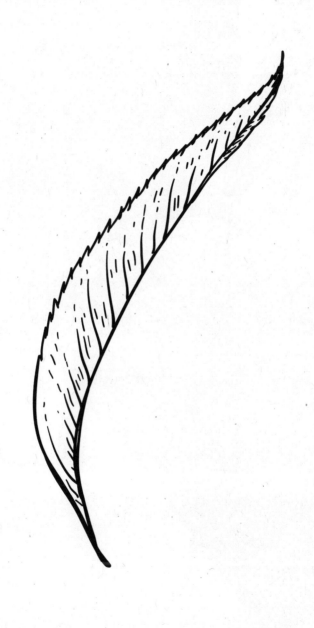

CHAPTER 6

Eating Animals (and Other Things)

My first experience with carnivorous plants was a disaster. One of my family members rightfully assumed I would be interested in growing a Venus fly trap, so they picked one up from a local nursery and presented it to me with joy in their eyes. I reciprocated by excitedly obsessing over the plant for two weeks straight. I spent hours staring at the thing, waiting for its sinister traps to reopen, only to trigger them again and again. Occasionally, I would find an ant or a worm to put in there, but my time with the plant was short-lived. Sadly, the fly trap did not come with any care instructions, and the internet was at least a decade away from finding its way into our house. After about two weeks in my care, it was dead. I later found out that watering it with tap water was what did it in (Venus fly traps and many other carnivores can't handle minerals in their water). My carelessness with the trapping mechanism probably didn't help, either. In more recent years, I have had much more luck in cultivating numerous carnivorous plant species.

As with all my obsessive ventures, I take great joy in learning the ins and outs of carnivorous plant cultivation. My need to get to know plants better opened a whole world of botanical discovery with carnivores. It didn't take me long to find out that the Venus fly trap is but one of many different species of carnivorous plant. It turns out, the world is full of these hungry little plants, and their means of catching prey are as varied as the species themselves. There are pitfall traps and sticky traps, snap traps and lobster pots, suction bladders, and even catapults. To capture prey, traps must be enticing. Often, this involves a bit of floral mimicry.

One of the best ways plants have evolved to capture prey is by disguising their traps as flowers. Using bright colors, floral scents, and even nectar, the business ends of carnivorous plants trick potential prey into thinking they are getting a meal instead of becoming one. Some carnivorous plants even cover their traps with special pigment cells that fluoresce in the ultraviolet wavelengths. Insects like bees and flies can see well into the ultraviolet portions of the electromagnetic spectrum, which is often utilized by flowers to produce a variety of patterns that attract pollinator attention. Dressed in UV pigments, the traps of these carnivores must seem pretty enticing. To our eyes, these pigments fluoresce blue under UV light, but to the eyes of a bee or fly, these areas would sport a unique patterning akin to some bizarre floral display that we can't fully appreciate with our limited visual acuity. Experiments have shown that when these pigments were masked, insects were far less likely to be caught, which strengthens the hypothesis that they function as lures. Such pigments have been found on the leaves of the Venus fly trap and surrounding the mouths of many pitcher plant traps. Even the fluids found within the pitchers of some species contain UV pigments.

The variety of lures aside, carnivory in plants is not a simple oddity of evolutionary history. Some form of carnivory has evolved in ten different plant families, and many of those origins are independent of one another. A carnivorous lifestyle can be very successful for plants. Once you get past the idea that plants can eat animals, the next question that usually comes to mind is why? What kind of selective forces in nature produce a carnivorous habit in a plant? The answer to this lies in where these plants tend to grow. Carnivorous plants hail from a variety of habitats around the world, but the one factor that unites these habitats is they tend to have poor soils lacking in nutrients like nitrogen and phosphorus. As varied as carnivorous plants are, their ability to consume animals is a means of supplying themselves with what the abiotic environment cannot. Thus, natural selection has honed these plants to be able to acquire nutrients via other organisms living in their environment.

It seems fitting that this chapter would follow the one on plant defenses as recent scientific discoveries have uncovered tantalizing insights into how carnivory in plants can evolve. Evolution doesn't necessarily require the appearance of novel genes or alleles. More often, evolution occurs via the retooling of genes that are already present within an organism. It would appear that at least some of the traits that allow plants to capture and consume animals got their start as defense traits. To understand how this works, let's take another look at the famed Venus fly trap.

It goes without saying that the Venus fly trap genome has been of interest to science for quite some time. However, despite many years of scientific scrutiny, no carnivore-specific genes have ever been identified. It's as if the carnivorous habits of this plant have no genetic basis. Of course, things are not that simple. A breakthrough was made when a team of scientists decided to look at chemical indicators of gene activity. When a gene is activated, it produces a chemical signal that can be easily traced. In doing so, scientists get a clearer picture of what is going on at a molecular level when an organism responds to its environment. To find out which genes were activated when a Venus fly trap captures a meal, scientists did what any of us would do—they fed the plant.

The "mouth" of a Venus fly trap is actually a highly modified leaf.

Instead of undiscovered carnivorous genes coming online, they found that the Venus fly trap has retooled a handful of genes that are used by many different plant species for defense. When the traps were stimulated by the insects stuck inside, transcripts that produce chitinase switched into high gear. For noncarnivorous plants, chitinase is usually produced in response to fungal infection. It just so happens that the polymer that makes up the cell membranes of fungi is chitin, the same polymer that makes up the exoskeleton of insects. Chitinase is an enzyme that specifically targets and breaks down molecules of chitin. Thus, the ability to digest insects has its origins in defending plants against fungi. This phenomenon is not unique to the Venus fly trap, either. Additional research has been done on various pitcher plants and sundews, and each has found similar results. The genes involved in consuming insects all seem to have been retooled from genes aimed at defending plants from fungal attacks.

Retooled is an understatement. There seems to be no end to the ways in which plants have managed to capture and consume other organisms. Among the most straight forward of these techniques are the pitfall traps. Pitfall traps, or pitchers, function in much the same way as the *Ceropegia* flowers we met back in Chapter 3. Regardless of the family of plants we are talking about, all pitchers form from highly modified leaves. The leaves change to develop in such a way that they create a deep cavity filled with fluid. In most instances, animals are lured in by bright colors, tantalizing scents, and even sweet nectar produced at the top or mouth of the pitcher. The cells surrounding the mouth of the pitcher are slippery, which makes it hard for anything paying a visit to keep a foothold. When an organism falls into the pitcher, slippery walls or even downward pointing hairs make escape nearly impossible. They are left struggling within the pitcher until they eventually drown and are digested.

Amazingly, four different plant families have evolved pitfall traps of some sort. We have the pitcher plants of North and South America in the family Sarraceniaceae, then there are the tropical pitcher plants in the family Nepentheaceae, which are native to much of Southeast Asia, Australia, and numerous Pacific islands. We also have a single oddball endemic to Australia that belongs to a family all its own, known as Cephalotaceae. Finally, there are a small handful of carnivorous bromeliads in the family Bromeliaceae. Each of these families are worth their own book, but they provide us with incredible examples of how effective the pitfall strategy can be.

The idea of a carnivorous bromeliad is almost laughable. They are relatives of the pineapple after all. However, a lot of species within this family have evolved to grow on trees and rocks, where water and nutrients are hard to come by. The so-called tank bromeliads have gotten around this by forming a bowl-like reservoir in the center of their leaves that can fill up with a surprising volume of liquid. Though obtaining water isn't much of an issue for tank bromeliads, getting the nutrients they need to grow and reproduce is often a bit more challenging. Most tank bromeliads rely on the gradual breakdown of debris within their tanks to slowly provide them with necessary elements like nitrogen and phosphorus, but at least two species of bromeliad in the genus *Brocchinia* don't bother leaving this to chance.

The thing about tank bromeliads is that they function like tiny oases wherever they grow. Everything from insects to monkeys will visit these plants to sip from their water reservoir. While *Brocchinia* have no use for animals as big as monkeys, they certainly benefit from insect visitors. Instead of splaying out their leaves like the spokes on a bike, *Brocchinia* grow them straight and tall. To me, they look like big, leafy test tubes. This surface of the leaves that form the inside are covered in a slick wax. These bromeliads even go as far as to lure prey in with sweet secretions produced near the tips of their leaves. Insects would be wise not to visit a *Brocchinia*, though. Their tall, columnar growth form coupled with slick, waxy leaves means some visitors inevitably fall into the liquid at the bottom. Once in the liquid, it is game-over. The insects drown, and digestive enzymes go to work extracting nutrients.

As odd as they are, *Brocchinia* provide us with a basic example of how pitfall traps can be used to capture animals. However, the more you look at variations on this carnivorous adaptation, the more complex it becomes. Take, for instance, the case of North America's most cold hardy pitcher plant, the purple pitcher plant (*Sarracenia purpurea*). It is easy to look at the pitchers of this plant and marvel at their potential killing power. Anything unfortunate enough to fall into its pitfall traps is sure to not make it out alive. However, if you were to spend some time investigating the insides of purple pitcher plant pitchers, you would find that they are teeming with life. Countless microorganisms and even a few invertebrates seem to be doing just fine within the digestive fluids. How could this be? Are these organisms immune to the digestive cocktail within the pitchers, or is something else going on in there?

Purple pitcher plants live in nutrient-poor bogs and supplement
their nutritional needs by trapping and digesting insects with
the help of some highly specialized symbiotic organisms.

Well, for starters, the role of digestive enzymes produced by the purple pitcher plant is a bit overstated. Active digestion in the purple pitcher plant seems to be a function of age. It is believed that the production of digestive enzymes is largely relegated to younger pitcher plants. The older the plants get, the less digestive enzymes they produce. As the plants mature and their pitchers become less hostile to life, a whole host of organisms find their way into the water within and, over time, a small but complex community develops. This community is referred to as an "inquiline community." Inquilines are animals that live in or around the dwelling of another organism. These relationships can be either parasitic or mutualistic, and both types of inquilines can be found within the pitchers of the purple pitcher plant. Bacteria and algae are the most abundant, but their presence allows for the survival of larger organisms farther up the food chain. Filter feeding rotifers and midge larvae soon find their way into the pitchers and begin feeding on the microorganisms within.

Within older pitchers, one can find the most specialized inquiline in the pitcher plant community, the larvae of the pitcher plant mosquito (*Wyeomyia smithii*). This mosquito breeds only within the water of the purple pitcher plant's pitchers! Pitcher plant mosquito larvae are the top predators within the pitcher community, feeding on other organisms that find their way either intentionally or unintentionally into the fluid inside. Before you freak out and think that pitcher plants are contributing to the mosquito populations that threaten us all summer, rest easy knowing that adult pitcher plant mosquitoes do not feed on humans. They instead feed on nectar, providing pollination for other bog dwelling plants.

So, what does the purple pitcher plant get in return for providing a home for all these organisms? Instead of digesting their own food, older pitcher plants are instead relying on all those bacteria and invertebrate to break down organic matter into usable nutrients for the plant. As the mosquito larvae, rotifers and other detritivores feed, they produce their own waste products, which also add to the mix of nutrients being taken up by the plant. Basically, as the inquiline community within each pitcher starts to provide all the added nutrients that the plant could ever need, older plants stop producing digestive enzymes, saving vital energy that can then be used for growth and reproduction. There are some organisms, however, that upset this system. The larvae of certain flesh flies will live in the pitcher, feeding and growing, until they are old enough to pupate. They will then chew a hole in the side of the pitcher to escape, effectively draining the plant of this entire life supporting ecosystem. Such is nature.

Not all pitcher plants have such a generalist diet. Some are far more specific and deadly. Consider the white-ringed pitcher plant (*Nepenthes albomarginata*) of Southeast Asia. This species has a taste for termites and does a remarkable job at turning the table on these notorious plant eaters. However, unlike flies or ants that are attracted to sweet nectar, termites have a different palate. With their penchant for feeding on plant materials, termites don't necessarily seem like the kind of insect a plant would want to attract, but the white-ringed pitcher plant has found a way to lure termites without becoming a meal itself.

The white ring around the rim of the white-ringed pitcher
plant is extremely attractive to termites.

As its species name "*albomarginata*" suggests, there is a white ring located around the mouth of each pitcher. The ring is made up of a dense coat of trichomes, which function as termite bait. Termites cannot resist them and once a scout finds a pitcher with a fresh ring of trichomes, the dinner bell is rung and hordes of termites flock to the plant. While many termites make off with a free meal, plenty more of them slip and fall into the pitcher. All of this happens in a span of a single evening. Once the ring is picked clean, the pitchers are no longer attractive to the termites. They go on their way, and the plant has its meal. Because of the social structure of these peculiar insects, the loss of these individuals is never high enough to be detrimental, thus the plant is free to maintain its extremely specific diet.

If you thought the white-ringed pitcher plant was oddly specific, wait until you meet the poop-eating pitcher plants of Borneo. At least two species stand out as fecal specialists. The first I would like to mention is Low's pitcher plant (*Nepenthes lowii*). Not only does this plant eat poop, its pitchers literally look like gaudy toilets dangling from the tips of its leaves. The mouth of each pitcher is extremely wide at the top, narrowing like a funnel to a constriction point roughly halfway down the length of the pitcher and then widening back out into a chamber at the bottom. Unlike the walls of insect-eating pitchers, the walls of these pitchers are not slippery at all, allowing most insects to crawl right back out. Indeed, insect capture rates are extremely low for this species. The key to its success lies in the lid attached to the back of the pitcher mouth. When a pitcher matures, this lid reflexes backward like a bristly backboard. The bristles secrete a foul-smelling, white goop that is said to be sweet to the taste.

The white goop appeals quite nicely to a species of tree shrew native to the mountains on which Low's pitcher plant grows. Shrews have been observed flocking to these pitchers to lap up the goop. To do so, they must position themselves in such a way that their back end is situated directly above the mouth of the pitcher. Close examination of the anatomy of the pitchers has revealed their dimensions to be a perfect fit for a perching tree shrew. As the shrew feeds on the goop, it also poops into the pitcher. Amazingly, tree shrew poop provides Low's pitcher plant with most of the nitrogen it needs.

Also found in Borneo is another species of pitcher plant known only as *Nepenthes hemsleyana*. Like many of its relatives, this pitcher plant grows as a vine, snaking its stem up into the canopy of its sturdier neighbors. Pitchers produced closer to the ground in this species are typical of what you would expect from a plant specialized in eating insects. They are scented, slippery, and full of viscous digestive fluids. They stand in stark contrast to pitchers produced further up in the canopy. The upper pitchers are considerably narrower, less scented, contain fewer digestive fluids, and are seven times less likely to capture insects. What's more, there is an obvious structure in the rear wall of the pitcher that is shaped like a parabolic dish. The reason that the upper pitchers of *N. hemsleyana* differ in form is because they just so happen to provide a wonderful roost for Hardwicke's woolly bats (*Kerivoula hardwickii*).

The narrowness of the upper pitchers coupled with the low levels of digestive fluids keeps bats safe from falling in and becoming a meal themselves. As a form of rent, the bats regularly defecate down into the pitcher as they roost, providing the plant with a nitrogen-rich meal of predigested bugs. It was found that the bats were also much more likely to roost in the pitchers of *N. hemsleyana* than they were in the other pitcher plants growing nearby. This is due to the frequency of the bat's echolocation, which is the highest of any bat species known to science. The frequency is likely an adaptation for the parabolic dish produced by the pitchers of *N. hemsleyana*. Studies show that it has strong echo-reflection for the frequency of the bat's sonar, which helps the bats find the pitchers among the dense vegetation of the Bornean rainforest.

Pitfall traps are fascinating botanical structures, but they largely rely on gravity for the initial stages of prey capture. Sticky traps take things to the next level by luring in their victims with the promise of a drink, only to snare them with viscous mucilage. Among the most famous adopters of this method of prey capture are the aptly named sundews in the genus *Drosera*. Though leaf shape varies from species to species, they are united in the production of stalked glands tipped in sticky mucilage. The combined effect of these glands gives the leaves a dewy appearance similar to what you see on your lawn at dawn and dusk, hence the common name. Sundews also differ from pitcher plants in that the leaves of many species are capable of movement. As hapless insects struggle to free themselves, they trigger more glands to close in around them. In some species, the entire leaf eventually curls around the insect, suffocating it in mucilage before the digestive enzymes are released and the insect is consumed.

The sticky leaves of a sundew ensnare unsuspecting prey.

How these plants achieve such surprisingly rapid movements in their leaves is still not fully understood but what scientists have discovered thus far is enough to keep you scratching your head for days. As the sticky glands sense the movement of a trapped insect, a chemical signal travels throughout the stalk of the gland, setting off a cascade of chemical reactions. First, the hormone auxin causes the cell to pump protons (H+) from within the cell out into the cell wall. This acidifies the cell wall, weakening its rigid structure. At the same time, a protein called expansin causes water to flow into some cells and not others, changing their sizes relative to one another. The change in size and sturdiness of different cells causes the leaf to bend inwards, trapping the struggling insect.

Sticky traps are found in more plants than the sundews. Another interesting example of sticky carnivores can be found growing in South Africa. The genus *Roridula* comprises two species sometimes referred to as fly bushes. Unlike most sundews which lay flat on the ground, fly bushes stand upright and branch, producing awl-shaped leaves covered in sticky glands. Their size and stickiness help these plants capture large prey like flies, moths, and wasps. However, unlike sundews, fly bushes don't excrete any digestive fluids. It's as if the plants only want to capture insects, not eat them. There must be some adaptive benefit to make up for the cost of production. A closer look at these plants revealed that, indeed, there is. It turns out that the fly bushes get a little help from another insect.

Living on fly bush plants are tiny capsid bugs that are covered in a special waxy substance which keeps them from getting stuck on the leaves. The bugs move freely about the sticky leaves, looking for trapped insects. When the insects are found, the capsid bugs impale them with their proboscis and suck them dry. As the capsid bugs feed, their droppings end up littering the fly bush leaves. This is how the plants obtain the nutrients they need to survive. Although the plants are not capable of actively digesting the insects, they can absorb the components of the capsid bug feces. Like the pitcher plants we met earlier, fly bushes are getting a little bit of fertilizer every time a capsid bug poops. By offering the capsid bugs a place to live and plenty of free, immobilized prey, the plant gets nitrogen-rich meals in return.

Among my favorite sticky plants are the butterworts (genus *Pinguicula*). The name sounds silly, but these plants do look like they have been wiped down in butter. Even the genus name *Pinguicula* roughly translates to "little greasy one." Butterworts can be found all over the world, but you are most likely to encounter the semi-tropical species native to Mexico. Like sundews, most butterworts form compact rosettes that hug the ground. They look more like succulents than they do carnivores, but a close examination of the leaves reveals they are covered in sticky glands. The leaves themselves are not capable of as much movement as those of a sundew, but when prey is captured, the leaf margins of some species will roll inward, helping to pool their digestive juices around the prey. Having grown a handful of Mexican butterworts, I can attest to the fact that their sticky traps are not very effective on larger insects. Occasionally, something the size of a moth will get its large wings helplessly stuck, but most of the time butterwort prey consists of smaller fare like fungus gnats and springtails. What these prey lack in size, they make up for in abundance. I have seen butterworts so laden with gnats and other small invertebrates that the leaves looked like someone had doused them in pepper.

The sticky leaves of butterworts are really good at catching smaller prey items.

Interestingly, insects aren't the only things on the menu for some butterworts. The diets of some species are composed of a surprising amount of plant material as well. Butterworts often grow in habitats dominated by pines and other wind pollinated plants. When these trees go into reproductive mode, so much pollen is released into the environment that it coats nearly every surface like ash from a volcano. This includes the leaves of resident butterworts. Pollen is high in protein, and many butterworts are capable of digesting pollen in addition to insects. Research indicates that the nutritional boost provided by all that pollen goes a long way in helping butterworts to grow, flower, and successfully reproduce. So, should we be calling these species omnivores instead of carnivores? How about we look at two additional cases of carnivorous plants expanding their diets before we decide.

Bladderworts (genus *Utricularia*) are a massive group of carnivorous plants. With something like 230 species spread around every continent on the globe except Antarctica, generalizing about their habits is not something I care to do. However, the most familiar of the bladderworts to those of us living in the northern hemisphere are the aquatic species that make their living floating in still bodies of freshwater. Don't let their size fool you, bladderwort traps are deadly. Each trap consists of a hollow bladder capped by a small lid. Whenever something small makes the mistake of coming too close to one of their trigger hairs, the lid snaps open, sucking the unlucky prey inside. This happens so fast that it can only be seen with the help of high-speed cameras. Once inside, the lid closes, making escape impossible, and the plant goes to work digesting its meal. Most prey consists of small invertebrates and the occasional baby fish;

however, more than animals wind up in those traps. The contents of aquatic bladderwort traps often contain a surprising amount of plant material as well. Most of this consists of single celled algae. Is it possible that at least some aquatic bladderworts also gain nutrients from all of that "vegetable" matter?

It does appear that algae are broken down within the traps, suggesting that the bladderworts are actively digesting them. Evidence of a nutritive benefit from algae digestion is mixed, however. Some studies have found that the bladderworts don't appear to benefit at all from the breakdown of algae within their traps, whereas others have found that bladderworts may obtain certain nutrients from digesting different types of algae. The benefit of trapping algae likely depends on the habitats in which bladderworts are growing. Plants living in more acidic water have been shown to capture far more algae than plants in more neutral or alkaline water, and pH seems to be the key. Numerically speaking, there are far less zooplankton than algae living in acidic water, which means algae is more likely to end up in the bladders. It could be that the benefits of digesting algae are greater for plants living in places where little zooplankton is available.

Not entirely convincing, but bladderworts go to show you that different strategies can work in different habitats. For a far more convincing example of a change in diet, let's travel back to Southeast Asia to look at a pitcher plant that is very serious about getting more vegetable matter in its diet. A rather widespread species, *Nepenthes ampullaria* differs from its carnivorous cousins in a multitude of ways. For starters, its pitchers are oddly shaped. Resembling an urn, they sit in dense clusters all over the jungle floor. Unlike other *Nepenthes*, the pitchers have only a small, vestigial lid with no nectar glands. Also, the slippery, waxy surface that normally coats the inside of most *Nepenthes* pitchers is absent. All these traits are clues to the unique way in which this species has evolved to acquire the nutrients it needs to survive.

Nepenthes ampullaria doesn't lure in and digest insects. Instead, it relies on leaf litter from the forest canopy above for most of its nutritional needs. Their urn-like shape, lack of a hood, and clustered growth enable the pitchers to capture considerable amounts of leaf litter. Because the pitchers are relatively long lived for a *Nepenthes*, lasting upwards of six months, they offer up a nice microhabitat for myriad other organisms. As with the purple pitcher plant, this inquiline community is the key to unlocking the nutrients in all that plant material.

In addition to countless microorganisms, scientists have found and identified upwards of sixty different species that rely on the pitchers for habitat. This list even includes a species of land crab and the larvae of one of the smallest frogs in the world. To make their pitchers more inviting for life, the plant actively manipulates its pitcher fluid, making it far less acidic than that of other *Nepenthes*. As the inquiline community breaks down the leaf litter, they release copious amounts of nitrogen-rich waste. The pitchers can then absorb this waste and begin to utilize it. Granted, this nitrogen is still coming from an animal source, but it all starts with leaf litter from neighboring plants. Pretty remarkable, if you ask me.

The open pitchers of Nepenthes ampullaria help this plant capture leaf litter from the canopy above, which is where most of its nutrients originate.

Pitfall traps, snap traps, bladder traps, and sticky traps; all of these are amazing adaptations that plants have evolved to catch and consume prey. But what about the lobster pot method I mentioned at the beginning of the chapter? Lobster pot certainly sounds like a strange way to describe any plant structure. For those unfamiliar, a lobster pot is a type of trap used to capture shellfish like lobsters, and it has a surprisingly simply design. Imagine a small cage made of wire and wood and wrapped in netting. At one end of the cage is a small, funnel-shaped opening that allows the lobster to enter. Once inside, escape is far more difficult as the funnel shape means that the exit hole is much smaller than the entrance hole. Amazingly, at least two unrelated species of carnivorous plant utilize a method very similar to this to obtain their meals.

One species is endemic to seasonally flooded areas of North America's coastal plain forests. It is a pitcher plant colloquially referred to as the parrot pitcher plant (*Sarracenia psittacina*). In my opinion, it is the most aberrant of all the *Sarracenia*. Instead of tall, lanky, upright pitchers, it produces a rosette of smaller pitchers that lie almost flush with the ground. Additionally, the leaf-like hood that covers the pitchers of its relatives has grown into a dome-like structure speckled with translucent patches and comes to a curved point near the front, hence the name parrot pitcher. Finally, the belly of each pitcher sports a leafy fin called an "ala" that runs the whole length of the tube.

The pitchers of the parrot pitcher plant are unique in that they lay flush with the ground and can capture prey both on land and underwater.

Its unique appearance is an adaptation to seasonal flooding and has changed the way in which this species captures prey. Unlike its relatives, the pitchers of the parrot pitcher plant do not function as pitfall traps. Lured in by their bright colors, insects gradually explore the traps thinking they might find a flower full of nectar. The fin-like ala directs these unsuspecting victims to the mouth of the pitcher. The translucent patches on the domed hood give the impression of multiple escape routes, enticing the visitor to crawl inside. Once inside, the insects become disoriented and cannot easily find the actual escape route. As they crawl farther into the pitcher, backward pointing hairs ensure that escape is impossible. Death is followed by digestion as the pitcher obtains yet another nutrient-rich meal. Insects aren't the only game in town for the parrot pitcher plant. Thanks to its prostrate habit, the parrot pitcher plant regularly finds itself underwater during floods. This would be bad news for most other pitchers as their upright position would allow whatever was inside to float out and away. Such is not the case for the parrot pitcher. Underwater, the pitchers function even more like a lobster trap. Everything from aquatic insects to tadpoles and fish can and do fall victim to this plant. Not even seasonal flooding can rob this unique pitcher plant of its meal ticket.

To find other examples of botanical lobster pots, one can expand their search range to semi-aquatic habitats in both Central and South America as well as the southern half of Africa. The group of plants you should be looking for are called corkscrew plants (genus *Genlisea*). All thirty or so species are small, producing tiny leaves positioned just above the saturated sandy soils in which they grow. Their flowers are elaborately shaped, strikingly colored, and hint at their familial relationship with the aforementioned bladderworts. Unlike more charismatic carnivorous plants, the meat-eating habit of this group is impossible to observe aboveground. To get a complete picture of their carnivorous lifestyle, we must look underground.

Corkscrew plants have no roots. Instead, they are anchored into the soil by truly bizarre, highly modified leaves. These leaves produce no chlorophyll and look absolutely nothing like a leaf. Instead, they form a hollow cylinder that corkscrews down into the permanently saturated soils. Along the length of each corkscrewed leaf runs a slit-like opening. Lining the mouth and inside of the chamber are backward pointing hairs. Like a lobster pot, animals can enter these slits with ease. Once inside, the only option critters have is to continue onward to their doom. Toward the end of the traps sits a chamber that contains almost no dissolved oxygen. Prey eventually succumb to suffocation and are digested. As you can probably imagine, the corkscrew plant menu largely consists of tiny, soil-dwelling organisms like protozoans and worms. It is a bit of a mystery exactly how or even if these plants actively attract their prey. Some researchers have found substances within the cylinders that are thought to act as chemical attractants, but, to date, this remains speculative. What amazes me the most is that their highly modified leaves have taken the place of roots, anchoring the plants to the ground and providing them with all the nutrients and water they need to survive.

Hopefully, by this point, I have convinced you that carnivorous plants are among the coolest plants in the world. There is something hauntingly beautiful about a plant eating an animal. It's as if evolution, having grown tired of the complacency of most animals, decided to turn the tables on them. Of course, evolution has no such agency. It simply works with what it has. The retooling of plant defense genes has resulted in unique and surprising adaptations for capturing and consuming prey. It is no wonder that carnivorous plants have captivated the imagination of scientists and non-scientists alike. From movies and plays to documentaries and books, few other groups of plants have managed to find their way into our culture in such varied ways. Even more exciting is that many of these species can, with some attention to detail, be grown in and around the home, provided you can source seeds and plants from ethical vendors. Like orchids, carnivorous plants are among the most heavily poached plants, and much of that poaching is done to supply the ever-growing demand for rare specimens. It is important that our love for these plants does not contribute to their decline or extinction in the wild.

But now, it's time to move on to another group of plants that have evolved even stranger means of obtaining the resources they need to survive. They do this by stealing from other plants or even fungi. A surprising variety of plants have evolved a parasitic lifestyle to one degree or another. In the next chapter, we will take a closer look at some of the strangest representatives of that group.

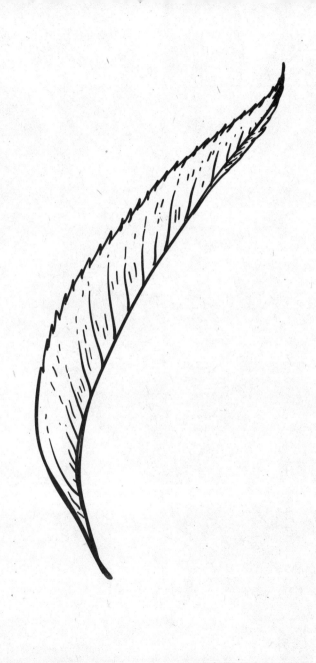

CHAPTER 7

Parasitic Plants

There I was plastered on a hillside somewhere in the southern tier of western New York. It was July and we were beating the heat by hanging out under the deep shade of a hemlock grove. The hillside we were perched on sloped abruptly down to a shale-bottom creek. The water slithered its way over smooth limestone until it plummeted forty feet over a ledge. The waterfall was not only stunning, it kicked up a fine mist that cooled the hot summer air.

I was still fairly new to plants at that point. I could barely recognize many of the major families that called this region home. However, my curiosity for botany was really starting to hit its stride. I was poking around the side of the trail looking to see what kinds of ferns we could identify when something odd caught my eye. I had no idea what it was, but something in my brain told me it was special. It wasn't very big, but it had an overall symmetry to it. Further investigation was in order.

I leaned into the slope and got down on its level. It quickly became clear that I was, in fact, looking at some sort of plant. Before me was a small brown stem topped with a spike of flowers. Each flower had six petals, the most prominent of which was a white lip covered in purple spots that stood in stark contrast against the rest of the plant. I was clueless. Up until this point, I had thought all plants were green. What I was staring at had no leaves, no greenery of any kind. Was this a flower stalk severed from something above? Could these be tree flowers? I gave it a gentle tug. It felt firmly rooted in the ground.

What was this bizarre plant? Was it native? I had so many questions. The plant was so distinctive that it didn't take me long to use my field guide and get it down to a genus. That was the day I met my first coralroot orchid, a member of the genus *Corallorhiza*. It was also my first encounter with a parasitic plant. The spotted coralroot had opened my mind to the lengths plants have gone to survive. I had always considered plants completely tied to the process of photosynthesis, but here before me was a plant that was challenging that idea. The sudden realization that there was no way this could be the only species of parasitic plant in this world was an interesting one to ponder. What realm had I just fallen into? What kinds of parasitic plants were out there?

I'll come back to the coralroots a bit later. First, I think we should back up a bit. Let me begin by setting the basis for what we are calling a parasite. To be considered a parasite *sensu stricto*, an organism must live off or in another organism, obtaining nourishment and protection, while the host organism receives no benefit in return. In other words, parasites are drains on their host. This "free-loading" lifestyle has earned parasites a nasty reputation. Yet, with nearly fifty percent of the lifeforms on our planet adopting a parasitic lifestyle in one form or another, to simply write parasites off as purely negative is an outlandish oversimplification. The simple truth of the matter is the deeper we dive into the world of parasitic organisms, the more we realize just how important they are in shaping life around us. Parasitic plants are no exception to this rule. They span an entire spectrum of evolutionary strategies that tailor fit them for a life of living off something else. That "something else" is usually another plant or even a fungus. Some parasitic plants still photosynthesize, whereas others, such as the spotted coralroot orchid I encountered, have lost that ability altogether.

The jump to parasitism was probably gradual for most lineages. To attempt to understand this process, we can begin by looking at plants growing around us today. Because evolution is not hierarchical and no one parasitic strategy is "the best," we have plenty of extant examples ranging those that can resort to parasitism when needed to those that can live only via parasitic means. The first parasitic plants likely started off as opportunists, parasitizing their neighbors when conditions required it. They probably did so via specialized structures called "haustoria." Haustoria are a conduit for parasitism for many plants. Haustoria start off as little projections emerging from the root of the parasite. Upon contacting either a root or a stem of a suitable host, the haustoria then penetrate the vascular tissues. Once inside, they act very much like roots, sucking up water and nutrients from their host.

For some parasitic plants, this relationship isn't obligate. They will tap into a host whenever times are tough. They can be very wasteful about it as well. In fact, some of the more skilled botanists can tell you if a plant might be parasitizing its neighbors simply by touching their lips to the leaves on a hot day. Whereas most of the nonparasitic plants will have closed their stomata to avoid drying out, parasitic plants keep on transpiring. Consequently, their leaves will feel much cooler to the touch on a hot summer day as the water they are sucking from their host evaporates from their leaves. It's not their water after all, so why bother shutting down? Such is the case for the many species of rhatany shrubs (genus *Krameria*). These beautiful plants parasitize a wide variety of other plants, stealing water and nutrients that are hard to come by in their arid homelands.

Their partial or "hemiparasitic" lifestyle gives them an undeniable advantage during the hottest months of the year. It is likely that the early days of parasitic plants were similar to the rhatany situation, with plants capable of living independently nonetheless figuring out how to get a leg up on survival by "cheating" the system a bit.

One of the best examples of what scientists think the early days of parasitism in plants may have looked like can be found in western Australia. It's a species of mistletoe locally referred to as the moojar or Christmas tree (*Nuytsia floribunda*) for its habit of blooming around that time of year. And what a display it makes. The entire canopy explodes with sprays of bright orange to yellow flowers. Each flower produces copious amounts of pollen and nectar, making them an important food source for resident pollinators. It also happens to be the largest species of mistletoe on Earth. Unlike the epiphytic mistletoes some of you may be familiar with, the moojar grows into a decent sized tree. Decked out in a full canopy of leaves, it is entirely capable of photosynthesizing on its own, producing all the carbohydrates it will ever need. Instead, it parasitizes other plants as a means of acquiring water and minerals.

The moojar is a root parasite. Its own roots fan out into the surrounding soil looking for those of neighboring plants. Amazingly, exploratory roots of individual moojar trees have been found upwards of 360 feet (110 m.) away from the tree. It doesn't seem to be too particular about its host, either. When an opportunity presents itself, the moojar takes it. When they find a potential host root, something incredible happens. They begin to form haustoria, which wrap around the host root, forming a collar-like structure that looks like a tiny doughnut. The collar gradually swells and a small horn forms on the inside. As a result of the swelling, the haustorial horn physically cuts into its victim's root. Once this cut is formed, the haustoria form balloon-like outgrowths which intrude into the xylem tissues of the root, forming the connection that allows the moojar to start stealing the water and minerals it needs.

Incredibly, the moojar is so opportunistic in its search for a host that its roots will wrap themselves around anything remotely root-like, living or not. A wide variety of inanimate objects have been found wrapped up in a haustorial embrace including dead twigs, rocks, and even electric cables. Its opportunistic parasitic nature appears to have left it open to exploring other, albeit dead end options. As disconcerting as this parasitic process may sound, this tree is not the enemy of surrounding vegetation. It is worth noting that the moojar extracts very little from any given host, so its impact is spread out among the surrounding vegetation, reducing its impact on any single plant.

The moojar is a great example of what the early days of parasitism may have looked like for plants. However, until we find fossil evidence, no one can say for certain how the parasitic ball gets rolling. What we do have available to us is a wide spectrum of parasitic strategies to study. From generalists like the moojar, we can move down the line to parasites that are a bit choosier about their hosts. By studying host-parasite dynamics, scientists are gaining wonderful insights into how these systems evolve. A fantastic example of this can be found in the Mojave and Sonoran deserts and involves yet another species of mistletoe.

The desert mistletoe (*Phoradendron californicum*) is not hard to spot, especially during the driest parts of the year when most of its host trees have shed their leaves. Unlike the moojar, the desert mistletoe is not free living, nor is it very large. It is a stem parasite that lives as an epiphyte, dangling from the branches of its host tree. It looks like a leafless, tangled mass of pendulous, photosynthetic stems that someone haphazardly tossed into the canopy. Like the moojar, the desert mistletoe is a type of hemiparasite, relying on its host tree for water and nutrients but is nonetheless capable of photosynthesis.

Luckily for the scientists who study it, we appear to be living during an important time in the desert mistletoe's evolutionary history. Some desert mistletoe populations seem to be heading toward a more specialized lifestyle based on their preferred host. Overall, desert mistletoes prefer leguminous trees like the palo verde (*Parkinsonia florida*), mesquites (genus *Prosopis*), and acacias (genus *Acacia*), but each population largely keeps to representatives of one type of tree. Scientists have been able to demonstrate that desert mistletoe preference isn't random, either. It starts at germination. For instance, when seeds collected from mistletoe growing on acacia trees were placed on paleo verde or mesquite, fewer of them germinated than if they were placed on another acacia. Though the exact mechanisms aren't clear at this point, evidence suggests that the success of desert mistletoe may be influenced by various hormone levels within the host tree, with isolated populations becoming more specialized on the chemistry of their specific host in that region.

The desert mistletoe is easy to spot, especially when
its host tree has dropped its leaves.

Scientists have also found evidence to indicate that populations of desert mistletoe growing on different host trees are reproductively isolated as well. Populations growing on mesquite trees flower significantly later than populations growing on acacia or palo verde. By flowering at different times of the year, their genes never have the chance to mix, thus increasing the differences between populations. Again, we are not entirely certain how the host tree may be influencing mistletoe flowering time; however, hormones and water availability are thought to play a role. Birds may also have a part to play in this drama, too. After pollination, the desert mistletoe produces copious amounts of bright red berries that birds find irresistible. Two birds in particular, the northern mockingbird and the Phainopepla, aggressively defend fruiting mistletoe shrubs within their territories. It is possible that these birds may be influencing which trees the seeds of the desert mistletoe end up on, further strengthening the specialization of each population.

The world of parasitic plants grows even stranger the more specialized they become. The more a parasitic plant relies on a specific host, the less work it must do by itself. Stealing water and nutrients are one thing, but there are parasitic plants that have abandoned photosynthesis altogether. If you have ever passed by a field or a section of hiking trail and witnessed what looked like a tangled pile of orange spaghetti tossed over the surrounding vegetation, then you have crossed paths with a highly specialized parasitic plant. Over a hundred species strong, members of the genus *Cuscuta* have conquered the globe and are absent only from the coldest regions of our planet. No doubt, they owe much of their success to us. *Cuscuta* are common parasites of

Parasitic Plants

numerous crops and find an easy living wherever we turn functioning ecosystems into monocultured ag fields.

Most often, these plants are referred to as dodder, but our disdain for their parasitic habits have earned them plenty of colorful nicknames like witch's hair, strangleweed, Devil's gut, and hellbine. Agricultural concerns aside, dodder are remarkable parasites. They produce no leaves, and most produce no chlorophyll, either. Roots are only ever produced by emerging seedlings, but are completely lost by the time the plant finds its host. Finding a host is critical. If a seedling cannot attach itself to another plant within ten days of germinating, it runs out of energy and dies. If you aimed a time lapse camera at a dodder seedling, you would see the emerging tendril frantically whipping about in a spiral like a tiny lasso. This is how dodder searches for a host. Its 360-degree rotation increases the chances of contacting the stem of another plant. Once it does, dodder doesn't waste any time.

The young dodder will wrap itself around the stem and begin to climb. At this point, it no longer needs access to soil and it discards its tiny roots. Wherever the stem makes good contact with its host, it will produce numerous haustoria that plug the dodder into the vascular tissues, giving it access to all the water, nutrients, and carbohydrates it will ever need. However, dodder can't just parasitize any nearby plant. They need to find a compatible host. As such, different species of dodder have developed a penchant for plant identification. Some dodder are especially fond of members of the tomato family, skipping over numerous different plants to get at their stems. This has puzzled scientists and led to many theories about how dodder discriminates among its neighbors. In 2006, a team of scientists discovered that dodder "sniffs out" its victims.

Of course, dodder don't have noses, but they do have the ability to sense volatile chemicals in the air. All plants release chemical compounds into the environment. Some are so pungent that even our less than stellar sense of smell can detect them. Others require more sophisticated senses to pick up on their subtle aromas. Each species of dodder is particularly sensitive to the gases given off by its preferred hosts. By narrowing down the different chemical compounds given off by dodder hosts, scientists have been able to identify which compounds most appealed to these parasites. They would rub those onto a piece of paper and watch as dodder seedlings literally trace the path along which the compounds were spread.

Amazingly, they also learned that certain chemicals can deter dodder. For instance, wheat plants release a compound called (Z)-3-hexenyl acetate that cause dodder to actively avoid the plant. It just so happens that (Z)-3-hexenyl acetate is also released by plants when they are under attack by herbivores like caterpillars. Because healthy, robust plants make better dodder hosts, it is likely that dodder also uses chemical cues to avoid plants that are sick or weak. However, the wheat in the experiment were perfectly healthy. Nothing was attacking them. Could it be that some plants have evolved a means to deter parasites like dodder? Could the wheat be feigning attack to appear less attractive? Most likely. By researching parasites, scientists are gaining incredible insights into the complexity of plant signaling and detection, and the story that is unfolding is far more complex than we ever imagined.

The bright orange stems of dodder parasitizing a
wild hydrangea (*Hydrangea arborescens*).

In recent years, it has also been discovered that dodder steal more than just water and nutrients from their hosts. They also steal genetic material. The movement of genetic material from the genome of one organism into the genome of another is called "horizontal gene transfer," and it is surprisingly common in nature. Microbes like bacteria do it all the time, and more and more, we are finding examples in complex multicellular organisms like plants. Scientists have identified over a hundred genes that have been added to the dodder genome via horizontal gene transfer. These genes come from a wide variety of host lineages, including representatives from the orders Malpighiales, Caryophyllales, Fabales, Malvales, Rosales, and Brassicales. Not only are many of these genes complete copies, they are actively transcribed by the dodder genome and are therefore functional. These include genes for the development of haustoria, genes for defense responses, and genes for amino acid metabolism. Scientists have also found an instance of a gene that codes for micro RNAs, which dodder will transfer back into the host plant, likely as a means of silencing host defense genes, allowing dodder to be a more successful parasite. We still don't know exactly how this process unfolds over time, nor if gene transfer from host to parasite is largely a one-way street. Still, evidence suggests that horizontal gene transfer is an important process among parasitic plants and may contribute to their success through evolutionary time.

When I hear stories of parasites like mistletoes and dodder, it makes me thankful that plants haven't yet evolved a means of parasitizing us. Sure, they have entirely taken over my life in other ways, but I can do without them physically invading my body. That said, parasitic plants have not restricted their efforts solely to members of their own kind.

Plants may be full of valuable water and nutrients, but they aren't the only game in town. There exists on this planet a multitude of plants that have taken to parasitizing fungi. In doing so, they have managed to trick one of the oldest and most important mutualistic relationships on our planet. We call these plants mycoheterotrophs, which is a fancy way of saying "fungus eaters," and they are among the most specialized. They are also far more common than you probably realize.

Just as we saw in the mistletoe examples, the ability to parasitize fungi probably didn't happen overnight. It was more likely a gradual process that began with both parties initially benefitting. Although we still have a lot to learn about mycoheterotrophic plants, we are lucky that a spectrum of plant-fungi partnerships exists today. Hints to the early days of fungal parasitism come from none other than my favorite plant family, the orchids. Would you believe me if I told you that every single orchid on this planet (provided it's not growing in a lab) starts out life as a parasite? Indeed, orchids aren't doting parents. They may produce a lot of seed, but they don't imbue them with much in the way of energy reserves to fuel germination and early growth. Orchid seeds need outside help if they are to survive.

Help comes in the form of mycorrhizal fungi. The fungi penetrate the seed coat and form a special kind of connection with the embryo inside. No one quite knows what spurs the initial hookup between these two different organisms. It's not as if the orchid embryo is exuding sugars or anything else the fungus might want. Also, neither organism has any foresight in terms of any future benefits the partnership might provide. Even in the best-case scenarios, it will be at least three to five years before the tiny orchid seedling has enough energy to produce its first leaves, allowing it to start providing the fungus with carbohydrates. No matter what the signal is, something keeps the fungus from completely overwhelming the orchid seed and digesting it. The two organisms enter into some sort of "agreement" with each other, and the tiny bundle of orchid cells goes from embryo to protocorm with the nutrients supplied to it by the fungus.

A protocorm looks nothing like a plant. It's not a stem, it's not a root; it's essentially an undifferentiated bundle of plant cells. Most protocorms live entirely out of sight and out of mind, sucking up whatever their resident fungus feeds them. Eventually, photosynthetic species will grow out of this phase and start repaying their fungus for all its hard work over the years, but along the way, a few different orchid lineages figured out how to back out on the mutualistic "agreement." Instead of growing photosynthetic tissues, these orchids live out their entire lives feeding on the fungus, never giving back anything in return. Their parasitic lifestyle can make these orchids hard to find. Since they never photosynthesize, they don't have to make an appearance aboveground until it's time to flower. Then and only then will you see them. The spotted coralroot orchid we met at the beginning of this chapter is a perfect example.

As you can imagine, this type of parasitic lifestyle comes with some unique adaptations, especially in the roots. The roots of mycoheterotrophic species don't really look like the roots of other plants. They form a strange mass adorned with funky protrusions jutting out in all directions. Stand on your head and look underground and you would see something that resembles a tiny undersea coral colony covered in fine threads. The technical term used to describe such roots is "coralloid," hence the common name "coralroot" and the generic name "*Corallorhiza*." Though coralloid roots are not restricted to coralroot orchids, these bizarre parasitic plants are among the most iconic plant to have evolved this anatomy.

There is still a lot of mystery surrounding what goes on inside coralloid roots, but what we do know is that the partnership starts out looking par for the course. The fungus grows into the root cells of the orchid and begin to form tiny, tree-like structures called pelotons. In nonparasitic situations, these structures facilitate in the exchange of water, nutrients, and carbohydrates between the fungus and its photosynthetic partner. In the roots of a mycoheterotroph, the pelotons are somehow broken down, turning what should be a mutual exchange into a one-way street that benefits only the orchid. We still have no idea if the fungus doesn't notice it's being parasitized or if, by some unknown mechanism, the plant is able to "convince" the fungus via chemical means to allow the parasitism to continue. All this mystery just makes me love these plants even more.

Once you get an eye for mycoheterotrophs, they become easier to spot. Their lack of photosynthetic machinery often means these parasites are adorned with colors we don't expect to see outside of floral structures. For instance, coralroot orchids are often dressed in pinks, red, and purple pigments. Even bright yellow species exist. The coralroots aren't alone in this, either. The mycoheterotrophic lifestyle has evolved in other flowering plant families as well. The blueberry family (Ericaceae) in particular has many parasitic representatives. In terms of colorful plants, mycoheterotrophic Ericads look as if they are trying to win "most eccentric" in some sort of botanical fashion contest.

Coralroot orchids like this stripped coralroot (*Corallorhiza striata*) make their living by parasitizing fungi and therefore do not need to produce leaves or chlorophyll.

There are plants like the ghostpipe (*Monotropa uniflora*) that truly earns this name with its eerie, ghostly pallor. Its lack of leaves, long, narrow stem, and single, bell-shaped flower curved toward the ground really do make this plant look like the ghost of my great-grandad's old pipe. Throughout its range, the ghostpipe has a multiflowered cousin called pinesap (*Monotropa hypopitys*) that comes in two different colors, depending on the time of year in which they flower. Populations that flower early in the summer tend to be golden yellow whereas populations that flower in early fall tend to be red. However, these species feel like mere palate cleansers compared to the visual vivacity presented by their cousins living in parts of western North America. There are species like the snow plant (*Sarcodes sanguinea*) whose fire engine red stems and flowers are so rich in color that it almost hurts your eyes. This plant's habit of blooming early means that it often displays its brightly colored reproductive structures against a backdrop of perfectly white snow, which is truly a sight to behold. However, in my opinion, the winner of "best dressed" parasitic ericad goes to the sugarstick plant (*Allotropa virgata*). This species combines the haunting white of the ghostpipe and the bright red of the snow plant into a single multiflowered stem that looks, for all intents and purposes, like a floral candy cane.

If strange-looking plants are your thing, then look no farther than the family Burmanniaceae. This family is comprised almost entirely of mycoheterotrophic species and has been experimenting (in an evolutionary sense) with both color and form. Most species in this family look more like they belong in the ocean than on land. The first few images I ever saw of this group gave me little indication that they were plants and not some sort of sea anemone. Most species are small, consisting of only some roots and the occasional flower, but that doesn't do them much justice. Do an image search of the family and you will be greeted by pictures of genera like *Thismia*, whose pea-sized flowers range in color from orange to red to blue to green and look either like tiny gas lanterns or hydra ringed with brightly colored tentacles. Then you have *Tiputinia foetida* which was only discovered in 2007 and is the sole member of its genus. This species only ever emerges as a single flower that lies flush with the ground and is doing its best impression of a starfish. Each petal is translucent brown, and they surround a bright yellow structure at the center that looks like a snowflake with its arms turned inwards. There seems to be no end to the diversity in shapes and colors in the Burmanniaceae family, and it's a shame that their small size and often remote nature means most people will never have the luck of crossing paths with one of these strange parasites.

Mycoheterotrophs are hyper specialized in their parasitic ways, but they nonetheless retain many features that allow you to recognize them as plants. They have roots, they have stems, and some even retain hints of leaves. However, there are parasitic plants out there that have done away with nearly all the physical trappings of the plant kingdom. This is because instead of living *off* other plants, these parasites live *in* them. Their relationship with their hosts has been taken to the absolute extreme. As you can imagine, this lifestyle necessitates its own unique set of physiological and morphological adaptations. To successfully invade a host to this degree, these parasites have reduced their physical body to something that moreso resembles a fungus than it does a plant. Slice into the tissues of an infected host and you would find a network of thread-like structures running throughout the vascular system. Those threads are the "body" of the plant.

Some of my favorite examples of this extreme parasitic lifestyle can again be found in the mistletoes. As you already learned, these plants can be very intimate with the hosts, with most members plugging themselves into stems and branches with haustoria. It shouldn't be a surprise, then, that a few species would take this to the next level. *Tristerix aphyllus* is a great example. Its hosts are cacti in the genus *Echinopsis* native to Columbia and Chile. You would never know a cactus had been infected until the mistletoe living within decides to flower. Bright red or yellow mistletoe flowers borne on branching stems will erupt from the succulent body of the cactus, making it look like the cactus is the one looking for love. However, a bit of scrutiny would reveal them to be radically different in form from a cactus bloom. Strangely enough, this mistletoe will occasionally produce small leaves on its flowering branches, reminding us of its photosynthetic past; however, it's believed that these leaves do little, if any, photosynthesizing.

Tristerix isn't alone in this habit, either. Across the Atlantic in Africa, there exists another species of parasitic mistletoe that lives out most of its life entirely within the vascular tissues of its host. Amazingly, this species specializes on a family of plants that, through convergent evolution, has come to resemble cacti in form and function. I am, of course, talking about cactus-like members of the spurge family (Euphorbiaceae) such as the African milk barrel (*Euphorbia horrida*).

The mistletoe that infects these plants is known to science as *Viscum minimum*. Here again, you will only ever see this mistletoe when it decides to flower. Small clusters of bright orange and green flowers erupt from the side of the *Euphorbia* and present themselves to pollinators. After these plants have been pollinated and their seed dispersed, they disappear back into their host until they have garnished enough energy to flower again.

The interesting thing about obligate parasites like these is that they are one-hundred percent reliant on the health and wellbeing of their host plant. If the host weakens and dies, it takes the parasite along with it. Flowering takes a lot of energy for plants, and because of this, one would think that the floral displays of such parasitic plants would remain on the smaller side of the spectrum. This is not always the case. Amazingly, the largest single flower in the world is produced by a parasitic plant that lives out its life within the vascular tissues of a tropical relative of grape vines. The giant corpse flower (*Rafflesia arnoldii*) hails from the jungles of southeastern Asia and only appears outside of its host vine when it's time to reproduce. Flowering begins when massive buds erupt from the vine. These gradually swell to resemble a large, gothic cabbage. Eventually the bud opens, revealing a flower of truly epic proportions.

At upwards of three feet (one m.) in diameter, the giant corpse flower (*Rafflesia arnoldii*) produces the largest single flower in the world.

A massive set of five rounded petals encircle a strange, bowl-like structure which houses the reproductive organs. The flower itself is the color of red meat and bespeckled in knobby white protuberances. The whole structure looks like an artistically arranged corpse, and that is exactly what the plant is going for. Upon opening, the flower adds to its macabre corpse mimicry the scent of rotting flesh. The combination of look and smell attracts its pollinators, carrion flies. Individual corpse flower blooms average around three feet (one m.) in diameter and can weigh as much as twenty-four pounds (eleven kg.); however, in January of 2020, a record-breaking flower was discovered that was nearly four feet in diameter, suggesting that this strange plant still has many surprises in store for scientists. If its massive flowers weren't strange enough, there is evidence that, like dodder, the corpse flower is also stealing genetic material from its host vine. No one is quite sure what use, if any, the stolen genes have for the corpse flower, but they may have something to do with maintaining their parasitic lifestyle.

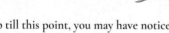

Up till this point, you may have noticed that all the parasitic plants I have mentioned are flowering plants. Indeed, the angiosperms have truly cornered the market on parasitism, but that doesn't mean there aren't nonflowering parasitic plants as well. Like orchids and Ericads, there are some nonvascular plants that also parasitize fungi. Scattered across the globe are liverworts in the aptly named genus *Cryptothallus* ("crypto" meaning hidden and "thallus" meaning sheet) that are entirely mycoheterotrophic. *Cryptothallus* liverworts have piqued the interest of enough botanists over the years to have earned the common name of ghostworts. Like the ghostpipe, they, too, are colored in ghostly shades of white. However, their nonvascular nature means the ghostworts must stick close to the ground where osmosis can readily supply them with the water they need to survive. In fact, most of their lives are spent almost completely buried in whatever substrate they are growing, making finding these strange parasites a difficult task. Their small size and relative rarity on the landscape likely helps ghostworts go unnoticed by the fungi they are parasitizing, but much more work needs to be done to better understand these dynamics.

At least one species of gymnosperm has also entered the realm of parasitism. Deep in the primeval forests of New Caledonia grows the only parasitic gymnosperm known to science, *Parasitaxus usta*. The sole member of its genus, *P. usta* is as strange and beautiful as it is mysterious. It is a member of a family of southern hemisphere conifers called Podocarpaceae, and it is beautifully colored in various shades of purple and wine. Although this bizarre gymnosperm does in fact produce chloroplasts, they are very small and the electron transport mechanisms that make photosynthesis possible no longer function. One of the strangest aspects of its morphology is that *P. usta* does not form any roots. This provided botanists the first clues that it may be a parasite. Further investigation has suggested that, like parasitic Ericads and orchids, *P. usta* utilizes a fungal intermediary to indirectly parasitize the roots of its only known host, another member of the Podocarpaceae family known as *Falcatifolium taxoides*.

Scientists have been able to demonstrate that the transfer of carbohydrates from host to parasite occurs entirely through this fungal connection, but *P. usta* also seems to obtain nitrogen and water via a direct connection to its host's xylem tissues. In this way, it is also like some mistletoes. As such, it not only can maintain a very high rate of growth, it can also produce cone crops year-round. To the best of my knowledge, no other parasitic plant on Earth adopts such a strange combination of strategies. Despite its unique status, much of the ecology of *P. usta* remains a complete mystery. For instance, it is entirely unknown how this parasitic gymnosperm becomes established on its host. The seeds aren't sticky, and to date, no seed dispersal mechanisms have been described. Perhaps it's all a matter of chance, which would explain why so few individuals have ever been found.

We are only just beginning to understand the true complexity of parasitic plants. Luckily, the technologies needed to peer into their secretive habits are becoming increasingly available to chronically underfunded scientific institutions. As scientists look closer at parasitic plants, they are uncovering amazing discoveries in chemistry, physiology, communication, DNA, and evolution. However, I don't think we can attribute one-hundred percent of their mystery to a lack of funding and technology. I think many of the blank pages in the story of parasitic plants can also be attributed to our collective disdain for parasites.

Humans, with our penchant for "rationalizing" the world around us, generally look at parasites with disgust. We demonize parasites as freeloaders, taking advantage of other hard-working organisms. Certainly, a big part of this distaste has its roots in our own evolutionary history. After all, parasites do sap their host and natural selection has rewarded our ancestors time and again for behaviors that help us avoid being infected. Also, from a more conscientious perspective, the thought of something living in or on us at our expense is enough to make our skin crawl. However, to completely write parasites off as a bane to all life would be a huge mistake on our part. More and more, we are realizing that parasites play an important role in ecosystem functioning and may even serve as indicators of ecosystem health. What's more, parasites are proving time and again to be a boon for biodiversity.

By knocking back the performance of their host, parasitic plants indirectly give non-host plants a competitive advantage, allowing them to establish in spots they would otherwise be crowded out. This often leads to an increase in overall plant diversity in that habitat. In effect, parasitic plants can level the playing field for other less competitive plant species. Parasitic plants can be so effective at giving other species a chance at survival that some gardeners and habitat restoration practitioners are starting to include parasitic plants into their seed mixes. By encouraging native parasitic plants to grow, significantly more plants can establish and thrive. For instance, here in the American Midwest, wherever the hemiparasitic wood betony (*Peducilaris canadensis*) gets its roots into competitive grasses like big bluestem (*Andropogon gerardii*) and Indiangrass (*Sorghastrum nutans*), they create what I like to call "betony balds," which are essentially areas of the prairie where grass growth is noticeably reduced, creating a small window for rare forbs like the royal catchfly (*Silene regia*) to grow and flourish.

In these uncertain times defined by wonton habitat destruction and biodiversity loss, we should be embracing all components of ecosystem health, including the parasites. No species on our planet operates in a vacuum. When it comes to protecting wild spaces and trying to put some of the pieces back together, we must act as holistically as possible. Biodiversity can mean a lot of different things depending on who you ask. Even the ways in which we quantify and try to understand biodiversity can vary. While it can be difficult to communicate the nuances of how the diversity of organisms within an environment scale up to ecosystem function, we can nonetheless be confident in one overarching conclusion: *biodiversity matters*. This is especially true for plants. As I stated earlier in this book, plants set the foundation for all other forms of life, especially in terrestrial systems. Without functioning native plant communities, everything else suffers. Unfortunately, discussions about ecosystems, conservation, and sustainability hardly ever mention plants. If they do, the topic is breezed over in favor of organisms deemed more charismatic like big cats or birds. The reality is, none of these organisms would exist if it were not for plants. A great deal of conservation efforts would benefit from a more plant-centric perspective. In this next and final chapter, I want to take the time to talk about the issues facing plants and what we can do to help. If there is one thing gardening in a suburban wasteland has taught me, it's that if you respect and foster native plant communities, life is sure to follow.

CHAPTER 8

The Problems Plants Face

Writing this book has been a bit of a struggle. I have tried to fill each chapter with as many of my favorite plants as possible, and while it has been a blast researching these species and learning more about their life here on Earth, a gloomy cloud has been hanging over my head the whole time. So many of the plants discussed in this book are not faring well in our increasingly modernized world. My struggle stems from the fact that every chapter could have easily devolved into a discussion about the plight of the species mentioned within. As I write this, an estimated forty percent of plants are at risk of extinction worldwide, and humans are to blame. Our population continues to grow, the standard of living is increasing in many countries, and human connectivity across the globe has never been stronger. As great as this may seem for us, all of it has come at great cost to the environment. Throw in climate change, and things only get worse.

I could just as easily have turned this into an eight-chapter rant against humanity, but I don't think that would have gotten us anywhere. As addictive as bad news is, focusing on it to the exclusion of all other messages is like sucking cold air over a cavity in your tooth. It's there and it hurts, but you aren't doing anything to change the situation. I understand that the living world is in trouble, but I want to do something about it rather than fall into a state of existential nihilism. That is why I decided to wait until the end to talk about the state of plants. All my experiences in podcasting and writing about plants has taught me an important lesson in communication: people do more when they are inspired than when they are depressed. Thus, my goal through most of this book has been to inspire you to look at plants in a new light and to appreciate them for the remarkable organisms that they are in the hope that you will begin to respect them enough to want to help. It is important to remember that as doom and gloom as the world may seem, something can (and must) be done. We can all have a stake in plant conservation, no matter where we live. We just have to try. So, in the interest of not spending the next few pages just complaining about the state of things, I also want to provide you with some examples of how you can get involved in fixing our world.

Habitat destruction is the leading cause of species extinctions. Without a place to live and breed, what does an organism have? When you learn of environmental destruction in the news or in blogs, you are usually presented with images of a clear-cut rainforest or entire hillsides engulfed in flames in areas like California or Australia. Certainly, these are examples of environmental destruction at its worst. However, as important as these images are for us to see, they give viewers the idea that environmental destruction is happening far away or is too large for any individual to tackle. How could a small housing development plopped into the middle of a rich cove forest somewhere in the Appalachian Mountains ever compare to the scorched Earth we witnessed in Australia or the hectares of palm oil plantations that are replacing biodiverse rainforests in Southeast Asia? By focusing only on the worst of the worst, we are ignoring all the environmental degradation and destruction that is going on around us all the time. Every new housing development that replaces a forest or remnant prairie, every new road the forest service cuts into a hillside, every new acre of soy and corn planted is eliminating habitat in the process.

Even just saying "habitat destruction" smooths over the nuances of what is really going on. People talk about habitat like it's an address of a house that has gone missing, but in reality, habitat starts with plants. Simply put, plants *are* habitat. Take pandas, for example. Pandas aren't endangered because we have hunted them to the brink of extinction. They are endangered because humans have cut and fragmented the bamboo forests that once covered huge swaths of south-central China to mere fragments of their former glory. Bamboo, as you know, not only comprises the backbone of panda habitat, it is also their main food source. Bamboo are essentially giant grasses, and as is typical with grasses, they can reproduce via underground rhizomes, forming dense stands of clones. Entire forests can be made up of the clones of only a few individuals. The bamboo that pandas require also experience a mass flowering event every sixty to a hundred years, with entire bamboo forests flowering all at once. After flowering and setting seed, the bamboo dies, and entire forests are lost in only a matter of weeks. Before humans fragmented their habitat, giant pandas had no trouble dealing with mass bamboo die offs. They just migrated to a new bamboo forest. They cannot do that anymore. When a bamboo forest flowers and dies, pandas in that area have nowhere to go, and they starve to death. Because of this, pandas now occupy a mere fraction of their former range. Despite considerable success in the captive breeding of pandas, there is not enough habitat left to support their recovery in the wild.

Pandas are lucky in that they captured our imagination and have garnered millions of dollars' worth of support. Most organisms are not so fortunate. This pattern repeats itself time and again across all branches of life. Few of us take any notice of it. As we discussed in

Chapter 1 with the loss of lupine and the Karner blue butterfly, even small critters like insects are not immune to the loss of the plants they need to survive. Here in the American Midwest, humans have reduced nearly 170 million acres of tallgrass prairie to less than four percent of its former glory, all in our quest to grow more corn and soybeans to feed cattle and sell to overseas markets. The descendants of the plants that once called those 170 million acres home have since had to eke out an existence within what few unplowed areas remain. Tiny pockets of remnant prairies can still be found in old cemeteries and along railways, but a lack of fire, too much mowing, and careless spraying of herbicide is making short work of those, too. When these plants disappear, so, too, do all the organisms that rely on them for food, shelter, and a place to breed. Such is the case for insects like the Eryngium root borer (*Papaipema eryngii*), a tiny brown moth that can only live where rattlesnake master (*Eryngium yuccifolium*) grows.

Rattlesnake master is a beautiful member of the carrot family that doesn't respond well to human disturbance. As soon as humans sink a plow into a prairie or unleash hordes of cattle upon it for grazing, rattlesnake master gradually disappears. This is bad news for the moth as its caterpillars can only feed on the interior tissues of rattlesnake master stems. When we remove rattlesnake master from the landscape, we are also removing a moth that depends on it. Now, imagine scenarios like this occurring wherever plants grow, which is damn near everywhere. Because plants set the foundation upon which all other organisms, from microbes and fungi to insects and birds, depend, destroying plant communities causes disastrous ripples that reverberate throughout the entire biosphere of our planet.

I often hear people ask questions like "it's just a bug, what purpose could it have had?" or "it's just a flower, why is it important?" and as infuriating as those questions can be, you must take a step back and realize where people with such ideas are coming from. As kids, many of us were taught that there is a hierarchy among living things, with humans taking the exalted rank at the top. We learn that everything that falls below us sits on rungs of diminishing importance. Unfortunately, plants with their unseeing, unhearing, unfeeling ways of life usually occupy the lowest rung of importance in our society. This style of thinking is deeply rooted in most religions, and whether intentional or not, it is spoon-fed to us from day one, greatly distorting our view of the natural world. It is also deeply untrue. Biodiversity matters, and we need plants to maintain it. I often hear the metaphor that nature is like an analog clock and all the animals are the teeth on the gears that make that clock work. The more teeth you lose, the more you reduce the function of that gear. The clock will gradually start to malfunction until one day, so many teeth are lost that it ticks its last tock. Well, if animals are the teeth, then plants are the gears on which the teeth sit. I'm no horologist, but I know that an analog clock can't work without gears.

With habitat destruction also comes invasive species. Non-native plants and animals are second only to loss of habitat in driving extinctions across the globe. People will argue tooth and nail over this, but you can't hide from the data. When you move an organism to a place where it did not evolve, or even give a native species an advantage it never had before, there is no telling how it is going to behave. Scientists have spent decades trying to figure out what makes something invasive in some areas and not others, but there is no single smoking gun. There are simply too many possible explanations to narrow it down to something tangible. For plants like garlic mustard that come equipped with chemical weaponry, arriving on a new continent with a favorable climate was all it took to start the invasion. For plants like buffelgrass (*Cenchrus ciliaris*), which was introduced into southwestern North America for cattle feed and erosion control, success comes from its ability to steal water and foster fire in a landscape that has not evolved to cope with burns. In places like the Sonoran Desert, buffelgrass moves in quickly and covers ground that would be bare otherwise. As it fills in around cacti, ocotillo, and creosote, it creates a massive fuel layer. All it takes is a single spark to set the grass ablaze, and the resulting wildfires scorch everything in their path. Unlike the native plants, buffelgrass quickly recovers from fire. When it does, it fills in the spots left vacant by dead native vegetation. Buffelgrass is literally changing the ecology of an entire ecosystem to its benefit, and it is by no means the only invasive to do this. As a species, we need to take invasive species seriously but plan our efforts wisely.

Another cultural issue plaguing plants is their utility. Since our earliest days on the African savanna, humans have used plants for food, medicine, fuel, and building materials. Although this is a fact that is worth celebrating, modern culture has done a great deal to pervert this mindset, making it seem as if the only interesting plants are those that are useful in some way. Growing up, the only messages I heard about plants involved their purported medicinal or culinary value. "Save the rainforest because we may find medicine!" dominated environmental talking points but, to date, I have seen no evidence of this message's efficacy. Similarly, older naturalists would take us out into the woods and wax poetic about the supposed healing qualities of a plant or recount stories of digging their favorite herbs and then lament that such plants are so hard to find these days. Ask them about the ecology of a plant, why its flowers have a certain shape, or what kinds of insects it supports, and you were met with blank stares of amusement. Local nature education groups would put on endless programs about the breeding habits of frogs, the migratory patterns of birds, or which mammals hibernate and which do not. When they did host plant programs, the topics were focused on foraging, herbalism, or making maple syrup. It's as if plants were put here by some divine force just so that we could utilize them. Yet, plants have been on land for over 450 million years and humans only about 300,000 years (give or take). By only discussing plants in the context of their utility, we are therefore only focusing on 0.07 percent of what makes plants so interesting.

Of course, plants are useful to us, and that should be celebrated. However, focusing purely on what plants can do for us sets a dangerous precedent. It teaches people that plants exist for our benefit and extracting them from the environment is not only a great way to connect with nature, but the whole reason they are there in the first place. I have seen firsthand the rampant disregard people can have for plants, no doubt fueled by this mindset. Foraging classes are on the rise, and every talk I give or article I write usually ends with people asking if any of these plants are edible. Granted, I would love to see our species rely less on industrial forms of agriculture, but acting like nature can support our modern needs is outlandish at best and more often a danger to ecosystem health. The Earth's natural areas are shrinking by orders of magnitude every year, and we simply can't afford to encourage cohorts of people to treat them as if it were the bulk bin at the grocery store.

Disregard for plants as organisms is a compounding issue that can hurt entire ecosystems. I remember one patch of forest near my childhood home that boasted a rich understory of ramps (*Allium tricoccum*) and wild ginger (*Asarum canadense*). One weekend, a group of foragers found that spot and harvested considerable quantities of both plants. This land was not open for extraction but extract they did. Where a carpet of ramp and ginger leaves once covered the ground, these poachers left nothing but churned up soil in their wake. The plants were harvested, roots and all. Not only did they not grow back, the following year, that spot was overrun with garlic mustard and dame's rocket, two invasive mustards whose seeds germinate far better in disturbed soils. What appeared to be a nice day of foraging for free food turned into an invasion hotspot that now threatens the existence of all the other species that grow nearby. Granted, not all foraging is poaching, but scenes like this play out all too often because plants are not afforded the same status as their animal counterparts.

Laws in the United States do almost nothing to protect plants. Permits are regularly issued for harvesting valuable species like ginseng (*Panax quinquefolius*) on public lands, but they are extremely hard to regulate. The situation is even worse on private land, where all you need is landowner permission to collect any plant you want. Shoot a bald eagle or hunt black bear out of season and you are looking at hefty fines and even jail time (rightly so). Take the last few individuals of an endangered plant, and barely anyone will notice, let alone care. Sure, you hear of the occasional person getting caught and fined for poaching ginseng in heavily regulated places like Great Smoky Mountains National Park, but national parks comprise a mere fraction of the natural areas in this country. The majority of plant habitat occurs on private lands where nefarious actions all too often occur out of sight and out of mind.

Illegal trafficking of plants is big business, but it rarely gets any attention. When we think of poaching, our minds naturally drift to animal products like ivory, shark fins, rhino horns, or pangolin scales. But would you believe me if I told you that trafficking of these products doesn't even hold a candle to poaching of plants like rosewood trees? Indeed, rosewood trees (genus *Dalbergia*) are the most heavily poached wild commodity in the world, and it's all because of their wood. Its rich rosy hue and pleasant scent have made rosewood extremely valuable, especially in Eastern markets, and people will pay outlandish prices to turn it into fancy pieces of furniture. This has led to rampant illegal logging in places like Guatemala, Africa, and Madagascar. Trees are felled at alarming rates, all to feed the ever-growing demand for rosewood. The situation is so bad that in the rainforests of Guatemala, scientists no longer consider these trees as living in populations. So many have been poached from forests that all that remains in most places are a few scattered individuals too small to be of any value. And again, rosewood poaching hurts more than the trees.

Because most of the logging is done illegally, little care is given to the fate of the surrounding ecosystem or the people who rely on it. The methods used to extract rosewood leave massive gaps in the forest, opening them up to the elements. In the forests of Madagascar, these gaps dry out quickly, subjecting the forest to increased risk from fires. Removal of the trees also harms the species that rely on them. Mature rosewood trees offer food and shelter to many of Madagascar's iconic wildlife. All poaching is infuriating but it just adds insult to injury to know that endangered species like rosewood trees and the ruffed lemur are being pushed closer to the brink of extinction because someone wanted a rose-colored chair to sit in.

Here in eastern North America, plants like ginseng and goldenseal (*Hydrastis canadensis*) are disappearing from our forests to fuel herbal markets both domestically and overseas. I once spoke with a ginseng biologist who spent years studying their population dynamics in the Appalachian Mountains. He recounted numerous situations in which the plants he was studying would disappear overnight with only a gaping hole in the ground as evidence of the poaching that occurred. Millions of individuals of these species are poached every year. Their numbers are growing smaller and smaller to the point that both are considered either threatened or endangered throughout much of their historical range. And their story is not unique. Whereas ginseng and goldenseal have managed to attract some attention and concern, other species are not so lucky. Plants like devil's bit (*Chamaelirium luteum*), ghost pipe (*Monotropa uniflora*), and black cohosh (*Actaea racemosa*) are dug from forests all the time and no one bats an eye.

Poaching is not limited to species of medical or culinary value. Plenty of plants are ripped out of the wild purely because they are beautiful or rare. Never underestimate the human urge to collect things. I, myself, am an avid plant collector. My obsession with gardening has me constantly looking for new and interesting species to cultivate. However, I draw a hard line at taking any plant from the wild. Unless I have permission to collect seeds, all the plants I grow are sourced from responsible friends or reputable nurseries. The same can't be said for some collectors. People will go to extraordinary lengths to get their hands on highly coveted species. Countless botanists have told me some version of a story that goes like this: "I used to lead nature hikes to spots with lots of orchids, but I had to stop because each year, someone would go back to that very spot and dig a bunch of them up."

Orchids are among the most heavily poached plants on our planet. Their beauty and mystique are all too often their downfall. The rarer an orchid is, the more people want it. The most unfortunate aspect of orchid poaching is that, at least for the terrestrial species, transplanting them is usually a death sentence. As we have learned, orchids require fungal mutualists to germinate and grow. For so many species, this relationship carries on into maturity, with the orchid relying on the nutrients provided by the fungus for its entire life. When someone digs up an orchid, it's like ripping the wires out of a computer. All those fungal threads are broken, severing the life-giving connections that keep the plant running. Unless the plant ends up in conditions that are far too ideal for most garden settings, those fungal symbionts will never return, and the plant will die. Decades of growth can be lost in a single moment because someone wanted a rare, pretty flower to add to their collection.

For orchids like these *Cypripedium calceolus*, habitat loss and poaching are having disastrous impacts on their numbers in the wild.

Orchid poaching is so serious that some outlandish efforts are being made to protect certain species. Take, for example, the case of a lady's slipper orchid known as *Cypripedium calceolus*. This wonderful plant is native throughout parts of Europe and Asia, but its numbers are on the decline. Habitat destruction and invasive species are the biggest issues this orchid faces, but its growing rarity means poaching is taking a heavy toll as well. However, poaching of this orchid is not a recent occurrence. In places like the UK, orchid poaching really hit its stride in the Victorian era. Orchids were being ripped from the wild wantonly to feed the publics' demand for these incredible plants. While lady's slipper populations on the mainland were able to weather this storm to some degree, populations in the UK were not so lucky. For decades, it was feared to have been extirpated until a single plant was rediscovered in the 1930s. That plant is still growing today, but to ensure that greedy hands don't come digging, an armed guard is stationed nearby while it is in bloom.

Similarly, the cycad hobby has produced quite a market for these strange gymnosperms. Like precious stamps or rare coins, collectors are scrambling to get their hands on the rarest of the rare. Unfortunately, cycads are the most endangered group of organisms on our planet, a tragic statistic you never hear about. Like countless other plants, the cycad lineage has survived for millions of years on our planet, weathering extinction event after extinction event, only to fall victim to a naked ape with the need for land and a penchant for collecting. Not only are rare cycads being taken from the wild, even plants housed in botanical gardens are targets for theft. In recent years, the Kirstenbosch National Botanical Garden in Cape Town, South Africa has experienced two incidents in which thieves broke in and stole twenty-four of their cycads in the middle of the night. Most of these were individuals of a species known as the Albany cycad (*Encephalartos latifrons*), of which only eighty remain in the wild. The plants that were stolen from the garden were vital to the future of this critically endangered species, and now, they are gone. The garden has since done what it can to beef up security, but they can only do so much. Cycads may be faring worse than any animal in Africa, but their plight is never front-page news. The lack of attention and funding toward studying and conserving plants like cycads means that the black-market trade in these organisms flourishes. Without money to investigate plant crimes, we have no idea how large the problem truly is.

At the core of all these issues is habitat fragmentation. Wildlife crimes like plant poaching would be far less of an issue if we had large contiguous tracts of habitat to support healthy plant populations. Alas, humans like to spread out, conquer new territory, and gobble up natural resources in the process. This doesn't always result in complete destruction. Often, our activities leave us with highly fragmented patches of wild spaces with very little nature in between. If you take an ecology class of any kind, you will eventually learn about the idea of island biogeography. Originally put forth in 1967 by Robert MacArthur and E. O. Wilson, island biogeography is a way of understanding biodiversity as it relates to isolated natural communities. It was first applied to oceanic islands, but its ability to accurately predict how many species a chunk of land of any given size can support has seen its application expand to all types of isolated natural communities such as those found on mountain peaks or even patches of unique soil types. The fundamental takeaway of island biogeography is beautifully simple; biodiversity is a function of immigration and extinction. Immigration increases the number of species in an area while extinction reduces that number. Immigration rates usually outpace extinction rates on larger islands because more land can support more species. Alternatively, extinction rates are usually higher on smaller islands, resulting in fewer species. The same rules apply for the degree of isolation. The more isolated a chunk of land becomes, the harder it is for species to get there, lowering immigration rates and vice versa.

When we fragment habitats of any kind, we are creating isolated islands. The smaller we make those fragments, the fewer species they can support. The more isolated these fragments become, the harder it is for those dwindling populations of plants and animals alike to be recharged by new arrivals. Sadly, there is increasing evidence that even modestly sized patches of remnant wild spaces are gradually losing species over time. Without some form of habitat connecting them, the species within slowly die off and aren't replaced. This is especially true for smaller plants with limited seed dispersal capabilities. For instance, plants like *Trillium* utilize ants for seed dispersal, but the ants they partner with cannot survive outside of the humid understory of the forest for very long. As such, the likelihood of an ant carrying a *Trillium* seed from one small patch of forest across vast stretches of farm fields and lawns is so small that I would put money on it almost never happening. Thus, *Trillium* living in isolated patches of forest have only each other and their resident ants to keep their population going. Of course, *Trillium* can't conquer every inch of that forest. Also, over time, inbreeding becomes an issue. If all the *Trillium* in a single patch of forest are pollinating only each other year after year, eventually, all the offspring are going to be related to one another. The shallower a gene pool gets, the more vulnerable those individuals become.

Vast stretches of human development fragment wild
spaces, reducing biodiversity in the process.

You don't have to eliminate every member of a species to doom it to extinction. Often, all it takes is reducing its numbers to a point in which it can no longer function properly. Inbreeding is one way this happens. Though this is yet another area where generalizing is difficult, genetic diversity is widely considered an important component in the resiliency of natural systems. If there is high genetic diversity in a population, it increases the chances that at least some individuals have the genes needed to overcome onslaughts like pests, diseases, or the ever-increasing rate of climate change. That is what is so worrying for species like the ash trees (genus *Fraxinus*) of North America. Ravaged by the feeding habits of the emerald ash borer (*Agrilus planipennis*), an invasive beetled introduced from Asia, North America has already lost millions of ash trees. The scariest part is that all those trees were lost in a little over a decade. The first signs of emerald ash borer invasion were reported in Michigan back in 2002, and today, it can be found in thirty-five states and five Canadian provinces. Stopping its spread is extremely difficult. There is no telling if North America will have any surviving ash trees in the next decade. The only hope comes in the form of whispers of a few trees that can resist ash borer infestations. Hopefully, those trees are truly resistant because their genes could form the foundation of ash breeding programs that may one day restore these incredible trees to a shadow of their former glory.

Genetic diversity and habitat connectivity are extremely important in the context of climate change. Yes, climates have changed in the past, but changes on the scale of global climate historically have occurred over great expanses of time. With thousands of years and vast stretches of habitat to support them, plants and other organisms were much more likely to acclimate and adapt. Extinctions still occurred, but never before have they resulted from the fast-paced actions of a single species. Humans have changed things drastically ever since we came onto the scene, and the most severe of our changes have occurred in little over a century. The rate at which we have altered ecosystems and our climate is unprecedented. Changes are now measured in human lifetimes instead of millennia.

One of the biggest failings of communicating the issues surrounding climate change is that we treat it as something that *will* happen rather than something that *is* happening. Every week, I talk with people who are working extremely hard to understand our natural world and how plants are faring within it, and all of them mirror similar sentiments that things are already changing. One conversation that really struck me was with Dr. Ken Feely, an ecologist studying trees growing in tropical forests of South America. His research takes him to some of the most remote corners in the Andes Mountains to better understand how climate change is affecting areas with little human influence. Even in these remote jungles, Dr. Feely and his colleagues are documenting evidence of the effects of climate change in tropical tree communities. By monitoring long-term plots, they have found that species from warmer, lower elevation forests are beginning to migrate up into the mountains. They appear to be following their preferred climate zones.

You can picture the climate zones of any mountain range as a series of stacked layers or rings with the warmest zones at the bottom and the coolest at the top. As our climate continues to change, these zones are moving upward, forcing the species living at different elevations to either keep up or be pushed out by species moving up from below. That is exactly what is going on in the Andes. The upside for remote locations like these is that humans have done very little to change the surrounding forest. Plants can move up the mountain because there are no clear cuts, roads, or developments to get in their way. Such is not the case in more developed areas. Where I do my research in the southern Appalachian Mountains, plants are responding to climate change as well, but they must contend with a lot of human infrastructure in the process. Mining, farming, logging, and housing developments all present significant barriers to species trying to keep up with favorable conditions. This is especially true for herbaceous species that live in forest understories. The low dispersal capabilities of many species may spell disaster in the coming decades as they are faced with increasingly stressful or even novel climate conditions. If plants like these have any hope of survival, a lot of them are going to have to adapt in place, but we still know very little about which plants can do so and which cannot.

Here, again, genetic diversity is key. The ancestors of plants were no strangers to change. Perhaps locked away in the genetic code of some species are the blueprints for acclimating and adapting to change. However, each chunk of forest that is clear-cut, each mountain top that is blasted into oblivion to get at the coal within, each new acre of lawn that is seeded replaces habitat and the genetic diversity of all the plants that once grew there. If all that remains of a species are a handful of isolated populations, it won't take much disturbance to push them beyond their limits. If most of those populations are genetically homogenous, they are all equally susceptible to change.

As depressing as these issues are, we can all do something about them. Until the governments of the world get their collective heads out of their asses, citizen action is our best option. Think global but act local because local is where things can get done. The first and most important thing we must do is protect wild spaces. This doesn't mean supporting only parks. It means protecting that abandoned lot at the end of the street, that chunk of forest behind your grandparents' house, or the small strip of remnant prairie that runs along the train tracks near town. We must start looking at such places as habitat rather than spots in need of mowing or development. Sure, small chunks of land may not be suitable for wolves or mountain lions, but they certainly serve a function for other forms of life. A single white oak tree (*Quercus alba*) can provide food and shelter for hundreds of different insect species, which will provide food for hundreds of other animals like lizards and birds. Never underestimate the impact of even modest sized patches of habitat. In states like Illinois, small patches are often all that remains. They not only serve as a safe haven for plants and animals, they can provide us with the pieces we will need to start putting this complex puzzle back together.

If you know of a land conservancy organization in your hometown, support them in any way you can. This doesn't always mean dipping into your wallet. If you are like me and cash is limited, give these organizations your time, instead. Volunteer to help monitor properties and easements or tag along when they host invasive species removal programs. At the very least, help them spread the word. Share their posts online, tell your friends and family, get other people to volunteer with you. If you don't have a local land conservancy, then turn your support to other nature-oriented groups in your region like your local native plant society or even larger organizations like The Nature Conservancy or the Rainforest Trust. Organizations like these live and die by public support and, in my opinion, they are doing the most important work for the environment. Protecting a chunk of land from development ensures that, at the very least, life will remain there. It may need help to restore some of its ecological function, but I would much rather see a fallow field full of "weedy" plants than another store filled with plastic and other useless things.

Of course, protecting habitat isn't enough. We must try to restore it as well. Ecological restoration is still in its infancy as a science, but that shouldn't stop us from making moves. We don't have the time to wait for the incremental pace of scientific progress before we get our hands dirty. We need to be doing as much habitat restoration as possible while taking every opportunity to learn along the way. Mistakes will be made, for sure, but they will pale in comparison to not doing anything at all. We can't be paralyzed by the need for data; we must get out there and start generating it. The thing about habitat restoration is that it's a lot like gardening; certain things will work in some areas but not others. Species that thrive in one restoration may struggle in another. Such is life. It's complex, it's messy, it's difficult to understand, but we know we need it. Life begets more life. Nowhere was this made more apparent to me than in our small apartment yard in Illinois. Our landlords are incredible people in allowing us to garden all we want. "Do what you want with it as long as it looks as good as, if not better, than before" were their exact words. Nearly six years later, our borders are chock full of native prairie plants that we have grown from seed and traded with friends around town. Where there was only turf grass before, species like Arkansas ironweed (*Vernonia arkansana*), pale purple coneflower (*Echinacea pallida*), and prairie blazing star (*Liatris pycnostachya*) now provide us with endless amounts of joy each summer as their flowers paint the yard in a rich variety of color. Life has responded to our efforts, too.

We rarely saw butterflies like monarchs in our yard the first two years after we moved in. A yard full of turf grasses doesn't offer much for insects. Now, with milkweeds (genus *Asclepias*) in every corner, monarchs have a place to lay their eggs. Their caterpillars find ample places to pupate within the tangled mass of stems that our blue vine (*Cynanchum laeve*) creates. When they emerge as adults, they feed with relish on the nectar of our rough blazing stars (*Liatris aspera*). Most amazing to me was the appearance of the pipevine swallowtail butterflies a few years ago. The caterpillars of this butterfly can feed only on the leaves of pipevines (genus *Aristolochia*). Right after we moved in, I purchased a few woolly Dutchman's pipe (*Aristolochia tomentosa*) from a native plant grower in town, and they have since taken over the fence that surrounds the property. As far as I can tell, our plants are the only pipevines around for miles. We live in the middle of a suburban wasteland, and our neighbors are definitely not plant people. How these butterflies found our vines in a sea of lawn and concrete is beyond me, but it makes me happy to see their chubby little caterpillars munching away each summer. Without those plants, I don't know where they would find food.

Where there was once only grass, a diversity of native
prairie plants now supports a variety of life.

You don't need to have a yard or even access to a garden to take part in restoring life to the landscape. If you have a window in which you can hang a window box or a balcony or patio where you can put some pots, you can bring plants into your life. If you do, try to utilize as many native plants as possible. Native flora support native fauna. Imagine what a lifeline a few potted coneflowers or milkweeds could provide to insects and birds struggling in a landscape dominated by concrete and blacktop. Even from an anthropocentric perspective, if eating wild foods and harvesting wild medicines is your thing, try growing them yourself. You don't need to put more pressure on wild populations. Instead, you can alleviate some of the pressure by producing it yourself. Also, the more plants we propagate, the more they become available to others. Trade with family and friends; host a yard sale that includes plants from your garden. Collect seeds from your plants, store some and give away the others. Often, people are not limited by a lack of desire for native plants, they are limited by their availability.

One of the best things you can do for plants and the environment as a whole is get rid of as much lawn as possible. Maintaining a lawn is akin to sterilizing the landscape. The turf grasses and other non-natives that comprise most lawns serve almost zero ecological function, especially compared to a garden filled with species indigenous to your region. If you own a property with lawn, try replacing some of it with a garden. Even better, let some of it return to nature. Like gardening, "rewilding" a bit of your yard can be a guided process. You don't have to let it go on its own. Instead, encourage native plants to establish by letting them grow or spreading some seed. Target aggressive non-native plants for removal and sit back and enjoy all the wildlife that will eventually return to your yard to take advantage of the habitat you have created. If that isn't an option, at least be lazier in maintaining your yard. Research is now showing that even reducing your mowing schedule and raising the blade a few inches can make a big difference for native bees and butterflies.

Also, start making noise in your community. Don't be annoying or combative, of course, but if you live in an area ruled over by a homeowner's association, then infiltrate that group. Start by (calmly) introducing the idea that gardens are a good thing and native plants aren't weeds. Demonstrate to them the beauty of the natural world. Most importantly, speak to them on their level. Don't attack them with data or jargon that only appeals to those who think like you. Instead, appeal to their sensibilities. Show them that a native plant garden doesn't have to look like a vacant lot, tell them how much energy you saved on your heating bill all thanks to the shade cast by the trees on your property, prove to them that mowing a little less saves time and money and restores some peace and quiet to the neighborhood. If you don't have a homeowner's association to gradually overthrow, then reach out into the community. Start going to public meetings and expressing your thoughts on how local parks are managed, what kinds of trees get planted, or how spraying of roadsides is doing more harm than good. Trust me when I say that most of the people attending and speaking up at such meetings are not always the most environmentally conscious. That needs to change. Town hall meetings all over the country need more rational, science-based opinions that bring ecology into focus.

The point I am trying to make here is that each and every one of us has options. Getting mad and sharing articles or memes on the internet isn't going to lift us out of impending ecological disaster, but getting up and doing something can. We can't afford to spend all our time being keyboard warriors. We must take action, and we must start now. As my friends at the Southeastern Grasslands Initiative like to say, "Twenty-five years will be too late." Waiting for the next generation to grow up and fix our problems isn't going to work. Our world is changing too fast. Globalization and constant connectivity make us feel like all is doomed. It isn't. There is still so much nature left to be saved and so much more that can be restored. We need to remember that and act. Ecology is not an issue best left to nature nerds or scientists. It is something we all need to care more about because all aspects of our lives depend on it. Whether you are a social activist trying to improve the lives of marginalized groups, a political scientist worried about growing unrest in countries ravaged by drought and famine, or a parent who wants their child to have a healthy and prosperous future, we all need functioning ecosystems to support us. Without functioning ecosystems, we have nothing.

If there is one message I hope to have instilled by the end of this book, it's that plants are the foundation of functioning ecosystems. Plants open our otherwise finite planet by taking energy from our nearest star and using it to make food. If photosynthesis never evolved, there is no telling how drastically different life would be, if it could even be at all. Plants supply food to other forms of life, which go on to provide food to other forms of life, and so on and so on. My other goal in writing this book was to show you that you don't have to be dragged kicking and screaming into appreciating plants as organisms. Once you get past all the hokey folklore and mysticism surrounding them, you will find that plants are endlessly fascinating in how they make a living and interact with the world. Plants are fighting for survival just like the rest of life on Earth, and their sessile nature has forced them to evolve unique and bewildering ways of getting by long enough to reproduce. Their abundance and diversity on our planet mean that a plant enthusiast will never get bored of exploring what makes plants so special.

So, my last bit of advice to you, the reader, is to put this book down, find yourself a field guide to the plants of your region, and get outside. Learn how to identify the plants in your neighborhood. Learn what makes them different. Most importantly, learn their names. Plant names are the key to opening the door to discovery. With a name, you can learn how that plant functions, where it likes to grow, and what kinds of life it supports. Eventually, you will come to know plants like you know your friends. You will look forward to seeing them year after year. Your excitement will give you a greater awareness of the world around you. You will begin to understand how changes in the weather can affect their growth and reproduction. You will begin to appreciate the niches they occupy as well as how quickly things can change when humans encroach on their land. To connect with plants is to be in connection with Earth. Figuratively speaking, you will begin to put down your own roots in the place you live. You will be more in tune with life. In the words of Edward Abbey, "It is not enough to fight for the land; it is even more important to enjoy it. While you can. While it's still here."

Bibliography

Chapter 2

Cipollini, D., & Cipollini, K. (2016). A review of garlic mustard (Alliaria petiolata, Brassicaceae) as an allelopathic plant. *The Journal of the Torrey Botanical Society, 143*(4), 339–348.

Tang, G. D., Ou, J. H., Luo, Y. B., Zhuang, X. Y., & Liu, Z. J. (2014). A review of orchid pollination studies in China. *Journal of Systematics and Evolution, 52*(4), 411–422.

Temeles, E. J., & Rankin, A. G. (2000). Effect of the lower lip of Monarda didyma on pollen removal by hummingbirds. *Canadian Journal of Botany, 78*(9), 1164–1168.

Whitten, W. M. (1981). Pollination ecology of Monarda didyma, M. clinopodia, and hybrids (Lamiaceae) in the southern Appalachian Mountains. *American Journal of Botany, 68*(3), 435–442.

Zhang, H., Li, L., Liu, Z., & Luo, Y. (2010). The butterfly Pieris rapae resulting in the reproductive success of two transplanted orchids in a botanical garden. *Biodiversity Science, 18*(1), 11–18.

Chapter 3

Breitkopf, H., Onstein, R. E., Cafasso, D., Schlüter, P. M., & Cozzolino, S. (2015). Multiple shifts to different pollinators fuelled rapid diversification in sexually deceptive Ophrys orchids. *New Phytologist, 207*(2), 377–389.

Donaldson, J. S. (1997). Is there a floral parasite mutualism in cycad pollination? The pollination biology of Encephalartos villosus (Zamiaceae). *American Journal of Botany, 84*(10), 1398–1406.

Epps, M. J., Allison, S. E., & Wolfe, L. M. (2015). Reproduction in flame azalea (Rhododendron calendulaceum, Ericaceae): a rare case of insect wing pollination. *The American Naturalist, 186*(2), 294–301.

Fleming, T. H., Sahley, C. T., Holland, J. N., Nason, J. D., & Hamrick, J. L. (2001). Sonoran Desert columnar cacti and the evolution of generalized pollination systems. *Ecological Monographs, 71*(4), 511–530.

Gaskett, A. C., Winnick, C. G., & Herberstein, M. E. (2008). Orchid sexual deceit provokes ejaculation. *The American Naturalist, 171*(6), E206–E212.

Gigord, L. D., Macnair, M. R., & Smithson, A. (2001). Negative frequency-dependent selection maintains a dramatic flower color polymorphism in the rewardless orchid Dactylorhiza sambucina (L.) Soo. *Proceedings of the National Academy of Sciences, 98*(11), 6253–6255.

Hansen, D. M., Beer, K., & Müller, C. B. (2006). Mauritian coloured nectar no longer a mystery: a visual signal for lizard pollinators. *Biology Letters, 2*(2), 165–168.

Heiduk, A., Brake, I., von Tschirnhaus, M., Göhl, M., Jürgens, A., Johnson, S. D., ...& Dötterl, S. (2016). Ceropegia sandersonii mimics attacked honeybees to attract kleptoparasitic flies for pollination. *Current Biology, 26*(20), 2787–2793.

Huffman, J. M., & Werner, P. A. (2000). Restoration of Florida Pine Savanna: Flowering Response of Lilium catesbaei to Fire and. *Natural Areas Journal, 20*, 12–23.

Johnson, S. D., Pauw, A., & Midgley, J. (2001). Rodent pollination in the African lily Massonia depressa (Hyacinthaceae). *American Journal of Botany, 88*(10), 1768–1773.

Luo, S. X., Yao, G., Wang, Z., Zhang, D., & Hembry, D. H. (2017). A novel, enigmatic basal leafflower moth lineage pollinating a derived leafflower host illustrates the dynamics of host shifts, partner replacement, and apparent coadaptation in intimate mutualisms. *The American Naturalist, 189*(4), 422–435.

Olesen, J. M., & Valido, A. (2003). Lizards as pollinators and seed dispersers: an island phenomenon. *Trends in ecology & evolution, 18*(4), 177–181.

Rosenstiel, T. N., Shortlidge, E. E., Melnychenko, A. N., Pankow, J. F., & Eppley, S. M. (2012). Sex-specific volatile compounds influence microarthropod-mediated fertilization of moss. *Nature, 489*(7416), 431–433.

Schiestl, F. P. (2005). On the success of a swindle: pollination by deception in orchids. *Naturwissenschaften, 92*(6), 255–264.

Schneider, D., Wink, M., Sporer, F., & Lounibos, P. (2002). Cycads: their evolution, toxins, herbivores and insect pollinators. *Naturwissenschaften, 89*(7), 281–294.

Sérsic, A. N., & Cocucci, A. A. (1996). A remarkable case of ornithophily in Calceolaria: food bodies as rewards for a non-nectarivorous bird. *Botanica Acta, 109*(2), 172–176.

Simon, R., Holderied, M. W., Koch, C. U., & von Helversen, O. (2011). Floral acoustics: conspicuous echoes of a dish-shaped leaf attract bat pollinators. *Science, 333*(6042), 631–633.

Smith, J. C. (2009). *Pollination by New Zealand geckos* (Doctoral dissertation).

Suinyuy, T. N., Donaldson, J. S., & Johnson, S. D. (2009). Insect pollination in the African cycad Encephalartos friderici-guilielmi Lehm. *South African Journal of Botany, 75*(4), 682–688.

Terry, I., Tang, W., Taylor, A., Singh, R., Vovides, A., & Cibrián Jaramillo, A. (2012). An overview of cycad pollination studies.

Terry, L. I., Roemer, R. B., Walter, G. H., & Booth, D. (2014). Thrips' responses to thermogenic associated signals in a cycad pollination system: the interplay of temperature, light, humidity and cone volatiles. *Functional Ecology, 28*(4), 857–867.

von Helversen, D., & von Helversen, O. (1999). Acoustic guide in bat-pollinated flower. *Nature, 398*(6730), 759–760.

von Helversen, D., & von Helversen, O. (2003). Object recognition by echolocation: a nectar-feeding bat exploiting the flowers of a rain forest vine. *Journal of Comparative Physiology A, 189*(5), 327–336.

Chapter 4

Azuma, A., & Okuno, Y. (1987). Flight of a samara, Alsomitra macrocarpa. *Journal of Theoretical Biology*, 129(3), 263–274.

Bobich, E. G. (2005). Vegetative reproduction, population structure, and morphology of Cylindropuntia fulgida var. mamillata in a desert grassland. *International Journal of Plant Sciences*, 166(1), 97–104.

Bodley, J. H., & Benson, F. C. (1980). Stilt-Root Walking by an Iriateoid Palm in the Peruvian Amazon. *Biotropica*, 67–71.

Gerlach, G. (2011). The genus Coryanthes: a paradygm in ecology. *Lankesteriana*.

Horn, M. H., Correa, S. B., Parolin, P., Pollux, B. J. A., Anderson, J. T., Lucas, C., ...& Goulding, M. (2011). Seed dispersal by fishes in tropical and temperate fresh waters: the growing evidence. *Acta Oecologica*, 37(6), 561–577.

Hovenkamp, P. H., Van Der Ham, R. W., Van Uffelen, G. A., Van Hecke, M., Dijksman, J. A., & Star, W. (2009). Spore movement driven by the spore wall in an eusporangiate fern. *Grana*, 48(2), 122–127.

Jackson, M. B., Morrow, I. B., & Osborne, D. J. (1972). Abscission and dehiscence in the squirting cucumber, Ecballium elaterium. Regulation by ethylene. *Canadian Journal of Botany*, 50(7), 1465–1471.

Junghans, T., & Fischer, E. (2008). Aspects of dispersal in Cymbalaria muralis (Scrophulariaceae). *Botanische Jahrbücher*, 127(3), 289–298.

Karlin, E. F., & Andrus, R. E. (1995). The sphagna of Hawaii. *Bryologist*, 235–238.

Lewanzik, D., & Voigt, C. C. (2014). Artificial light puts ecosystem services of frugivorous bats at risk. *Journal of Applied Ecology*, *51*(2), 388–394.

Mallón, R., Barros, P., Luzardo, A., & González, M. L. (2007). Encapsulation of moss buds: an efficient method for the in vitro conservation and regeneration of the endangered moss Splachnum ampullaceum. *Plant Cell, Tissue and Organ Culture*, *88*(1), 41–49.

Rolena, A. J., Paetkau, M., Ross, K. A., Godfrey, D. V., Church, J. S., & Friedman, C. R. (2015). Thermogenesis-triggered seed dispersal in dwarf mistletoe. *Nature Communications*, *6*(1), 1–5.

Smith, B. W. (1950). Arachis hypogaea. Aerial flower and subterranean fruit. *American Journal of Botany*, 802–815.

Swaine, M. D., & Beer, T. (1977). Explosive seed dispersal in Hura crepitans L. (Euphorbiaceae). *New Phytologist*, *78*(3), 695–708.

Western, T. L. (2012). The sticky tale of seed coat mucilages: production, genetics, and role in seed germination and dispersal. *Seed Science Research*, *22*(1), 1.

Warren, R. J., Elliott, K. J., Giladi, I., King, J. R., & Bradford, M. A. (2019). Field experiments show contradictory short-and long-term myrmecochorous plant impacts on seed-dispersing ants. *Ecological Entomology*, *44*(1), 30–39.

Chapter 5

Eshbaugh, W. H. (1987). Plant-ant relationships and interactions—
Tillandsia and Crematogaster. *Proceedings of the Second Symposium
on the Botany of the Bahamas*, 7–11.

Frederickson, M. E., & Gordon, D. M. (2007). The devil to pay: a
cost of mutualism with Myrmelachista schumanni ants in 'devil's
gardens' is increased herbivory on Duroia hirsuta trees. *Proceedings
of the Royal Society B: Biological Sciences, 274*(1613), 1117–1123.

Gilliam, F. S. (2019). Response of herbaceous layer species to canopy
and soil variables in a central Appalachian hardwood forest
ecosystem. *Plant Ecology, 220*(12), 1131–1138.

Hewitt, R. E., & Menges, E. S. (2008). Allelopathic effects of
Ceratiola ericoides (Empetraceae) on germination and survival of
six Florida scrub species. *Plant Ecology, 198*(1), 47–59.

Huxley, C. R. (1978). The ant-plants Myrmecodia and Hydnophytum
(Rubiaceae), and the relationships between their morphology, ant
occupants, physiology and ecology. *New Phytologist, 80*(1), 231–268.

Koch, G. W., Sillett, S. C., Jennings, G. M., & Davis, S. D. (2004). The
limits to tree height. *Nature, 428*(6985), 851–854.

Maccracken, S. A., Miller, I. M., & Labandeira, C. C. (2019). Late
Cretaceous domatia reveal the antiquity of plant-mite mutualisms
in flowering plants. *Biology Letters, 15*(11), 20190657.

O'Dowd, D. J., & Willson, M. F. (1997). Leaf domatia and the
distribution and abundance of foliar mites in broadleaf deciduous
forest in Wisconsin. *American Midland Naturalist*, 337–348.

Ohse, B., Hammerbacher, A., Seele, C., Meldau, S., Reichelt, M., Ortmann, S., & Wirth, C. (2017). Salivary cues: simulated roe deer browsing induces systemic changes in phytohormones and defence chemistry in wild-grown maple and beech saplings. *Functional Ecology*, 31(2), 340–349.

Orrock, J., Connolly, B., & Kitchen, A. (2017). Induced defences in plants reduce herbivory by increasing cannibalism. *Nature Ecology & Evolution*, 1(8), 1205–1207.

Reeves, R. D., van der Ent, A., & Baker, A. J. (2018). Global distribution and ecology of hyperaccumulator plants. In *Agromining: farming for metals* (pp. 75–92). Springer, Cham.

Weber, R. A. (1891). RAPHIDES, THE CAUSE OF THE ACRIDITY OF CERTAIN PLANTS. *Journal of the American Chemical Society*, 13(7), 215–217.

Chapter 6

Adamec, L. (2007). Oxygen concentrations inside the traps of the carnivorous plants Utricularia and Genlisea (Lentibulariaceae). *Annals of Botany*, 100(4), 849–856.

Anderson, B. (2005). Adaptations to foliar absorption of faeces: a pathway in plant carnivory. *Annals of Botany*, 95(5), 757–761.

Bradshaw, W. E., & Creelman, R. A. (1984). Mutualism between the carnivorous purple pitcher plant and its inhabitants. *American Midland Naturalist*, 294–304.

Cheek, M. (1988). 99. Sarracenia psittaceina: Sarraceniaceae. *The Kew Magazine*, 5(2), 60–65.

Clarke, C. M., Bauer, U., Lee, C. I. C., Tuen, A. A., Rembold, K., & Moran, J. A. (2009). Tree shrew lavatories: a novel nitrogen sequestration strategy in a tropical pitcher plant. *Biology Letters*, 5(5), 632–635.

Fukushima, K., Fang, X., Alvarez-Ponce, D., Cai, H., Carretero-Paulet, L., Chen, C., ...& Hoshi, Y. (2017). Genome of the pitcher plant Cephalotus reveals genetic changes associated with carnivory. *Nature Ecology & Evolution*, 1(3), 1–9.

Grafe, T. U., Schöner, C. R., Kerth, G., Junaidi, A., & Schöner, M. G. (2011). A novel resource-service mutualism between bats and pitcher plants. *Biology Letters*, 7(3), 436–439.

Koller-Peroutka, M., Lendl, T., Watzka, M., & Adlassnig, W. (2015). Capture of algae promotes growth and propagation in aquatic Utricularia. *Annals of Botany*, 115(2), 227–236.

Kurup, R., Johnson, A. J., Sankar, S., Hussain, A. A., Kumar, C. S., & Sabulal, B. (2013). Fluorescent prey traps in carnivorous plants. *Plant Biology*, 15(3), 611–615.

Legendre, L. (2000). The genus Pinguicula L. (Lentibulariaceae): an overview. *Acta Botanica Gallica*, 147(1), 77–95.

Moran, J. A., Clarke, C. M., & Hawkins, B. J. (2003). From carnivore to detritivore? Isotopic evidence for leaf litter utilization by the tropical pitcher plant Nepenthes ampullaria. *International Journal of Plant Sciences*, 164(4), 635–639.

Moran, J. A., Merbach, M. A., Livingston, N. J., Clarke, C. M., & Booth, W. E. (2001). Termite prey specialization in the pitcher plant Nepenthes albomarginata—evidence from stable isotope analysis. *Annals of Botany*, 88(2), 307–311.

Płachno, B. J., Adamus, K., Faber, J., & Kozłowski, J. (2005). Feeding behaviour of carnivorous Genlisea plants in the laboratory. *Acta Botanica Gallica*, 152(2), 159–164.

Poppinga, S., Hartmeyer, S. R. H., Seidel, R., Masselter, T., Hartmeyer, I., & Speck, T. (2012). Catapulting tentacles in a sticky carnivorous plant. *PLOS One*, 7(9), e45735.

Schulze, W. X., Sanggaard, K. W., Kreuzer, I., Knudsen, A. D., Bemm, F., Thøgersen, I. B., ...& Escalante-Perez, M. (2012). The protein composition of the digestive fluid from the venus flytrap sheds light on prey digestion mechanisms. *Molecular & Cellular Proteomics*, 11(11), 1306–1319.

Williams, S. E., & Pickard, B. G. (1980). The role of action potentials in the control of capture movements of Drosera and Dionaea. In *Plant Growth Substances 1979* (pp. 470–480). Springer, Berlin, Heidelberg.

Chapter 7

Calladine, A., & Pate, J. S. (2000). Haustorial structure and functioning of the root hemiparastic tree Nuytsia floribunda (Labill.) R. Br. and water relationships with its hosts. *Annals of Botany*, 85(6), 723–731.

Gibson, C. C., & Watkinson, A. R. (1992). The role of the hemiparasitic annual Rhinanthus minor in determining grassland community structure. *Oecologia, 89*(1), 62–68.

Hynson, N. A., Madsen, T. P., Selosse, M. A., & Merckx, V. S. F. T. (2013). Mycoheterotrophy: the biology of plants living on fungi.

Mauseth, J. D., Montenegro, G., & Walckowiak, A. M. (1984). Studies of the holoparasite Tristerix aphyllus (Loranthaceae) infecting Trichocereus chilensis (Cactaceae). *Canadian Journal of Botany, 62*(4), 847–857.

Mauseth, J. D., & Rezaei, K. (2013). Morphogenesis in the Parasitic Plant Viscum minimum (Viscaceae) Is Highly Altered, Having Apical Meristems but Lacking Roots, Stems, and Leaves. *International Journal of Plant Sciences, 174*(5), 791–801.

Mescher, M. C., Runyon, J., & De Moraes, C. M. (2006). Plant host finding by parasitic plants: a new perspective on plant to plant communication. *Plant Signaling & Behavior, 1*(6), 284–286.

Overton, J. M. (1997). Host specialization and partial reproductive isolation in desert mistletoe (Phoradendron californicum). *The Southwestern Naturalist*, 201–209.

Sinclair, W. T., Mill, R. R., Gardner, M. F., Woltz, P., Jaffré, T., Preston, J., ...& Möller, M. (2002). Evolutionary relationships of the New Caledonian heterotrophic conifer, Parasitaxus usta (Podocarpaceae), inferred from chloroplast trnL-F intron/spacer and nuclear rDNA ITS2 sequences. *Plant Systematics and Evolution, 233*(1–2), 79–104.

Smith, D. (2000). The population dynamics and community ecology of root hemiparasitic plants. *The American Naturalist, 155*(1), 13–23.

Yang, Z., Wafula, E. K., Kim, G., Shahid, S., McNeal, J. R., Ralph, P. E., ...& Person, T. N. (2019). Convergent horizontal gene transfer and cross-talk of mobile nucleic acids in parasitic plants. *Nature Plants*, 5(9), 991–1001.

Wickett, N. J., & Goffinet, B. (2008). Origin and relationships of the myco-heterotrophic liverwort Cryptothallus mirabilis Malmb. (Metzgeriales, Marchantiophyta). *Botanical Journal of the Linnean Society*, 156(1), 1–12.

Chapter 8

Borer, E. R. Conservation Assessment for Eryngium Root Borer (Papaipema eryngii).

Burghardt, K. T., Tallamy, D. W., & Gregory Shriver, W. (2009). Impact of native plants on bird and butterfly biodiversity in suburban landscapes. *Conservation Biology*, 23(1), 219–224.

Fadrique, B., Baez, S., Duque, A., Malizia, A., Blundo, C., Carilla, J., ...& Malhi, Y. (2019). Widespread but heterogeneous responses of Andean forests to climate change (vol 564, pg. 207, 2018). *Nature*, 565(7741), E10–E10.

Fusco, E. J., Finn, J. T., Balch, J. K., Nagy, R. C., & Bradley, B. A. (2019). Invasive grasses increase fire occurrence and frequency across US ecoregions. *Proceedings of the National Academy of Sciences*, 116(47), 23594–23599.

Keeley, J. E., & Bond, W. J. (1999). Mast flowering and semelparity in bamboos: the bamboo fire cycle hypothesis. *The American Naturalist*, 154(3), 383–391.

Pokladnik, R. J. (2008). *Roots and remedies of ginseng poaching in central Appalachia* (Doctoral dissertation, Antioch University).

Randriamalala, H., & Liu, Z. (2010). Rosewood of Madagascar: Between democracy and conservation. *Madagascar Conservation & Development, 5*(1).

Sanders, S., & McGraw, J. B. (2005). Harvest recovery of goldenseal, Hydrastis canadensis L. *The American Midland Naturalist, 153*(1), 87–94.

Vardeman, E., & Runk, J. V. (2020). Panama's illegal rosewood logging boom from Dalbergia retusa. *Global Ecology and Conservation,* e01098.

Willis, K. J. (2017). *State of the World's Plants Report-2017.* Royal Botanic Gardens.

Acknowledgements

I would first and foremost like to thank my parents, whose constant love and support got me to where I am today. I would also like to thank Sara Johnson for always being there for me through thick and thin and believing in me even when I don't. Finally, I would like to thank all the scientists who put in the blood, sweat, and tears to understand and conserve plants and the habitats they comprise.

About the Author

Matt Candeias holds an MA in Community Ecology from SUNY Buffalo State and a PhD in ecology from the University of Illinois. Matt is also the host of the *In Defense of Plants Podcast*, an internationally recognized, weekly podcast that celebrates his love for the botanical world. Since its launch in 2015, the *In Defense of Plants Podcast* has consistently ranked in the top fifty science podcasts on all major podcast portals. *In Defense of Plants* articles have been published in a wide range of gardening and science-based publications, both online and in print. When he is not podcasting or writing about plants, Matt can usually be found hiking, botanizing, and practicing photography.

Mango Publishing, established in 2014, publishes an eclectic list of books by diverse authors—both new and established voices—on topics ranging from business, personal growth, women's empowerment, LGBTQ studies, health, and spirituality to history, popular culture, time management, decluttering, lifestyle, mental wellness, aging, and sustainable living. We were recently named 2019 *and* 2020's #1 fastest growing independent publisher by *Publishers Weekly*. Our success is driven by our main goal, which is to publish high quality books that will entertain readers as well as make a positive difference in their lives.

Our readers are our most important resource; we value your input, suggestions, and ideas. We'd love to hear from you—after all, we are publishing books for you!

Please stay in touch with us and follow us at:

> Facebook: Mango Publishing
> Twitter: @MangoPublishing
> Instagram: @MangoPublishing
> LinkedIn: Mango Publishing
> Pinterest: Mango Publishing
> Newsletter: mangopublishinggroup.com/newsletter

Join us on Mango's journey to reinvent publishing, one book at a time.